Generalized Functions, Volume 3

Theory of Differential Equations

GENERALIZED FUNCTIONS, VOLUME 3

THEORY OF DIFFERENTIAL EQUATIONS

I. M. GEL′FAND
G. E. SHILOV

TRANSLATED BY MEINHARD E. MAYER

AMS CHELSEA PUBLISHING
American Mathematical Society • Providence, Rhode Island

2010 *Mathematics Subject Classification.* Primary 46Fxx.

For additional information and updates on this book, visit
www.ams.org/bookpages/chel-379

Library of Congress Cataloging-in-Publication Data

Names: Gel′fand, I. M. (Izrail′ Moiseevich) | Shilov, G. E. (Georgiĭ Evgen′evich)
Title: Generalized functions / I. M. Gel′fand, G. E. Shilov ; translated by Eugene Saletan.
Other titles: Obobshchennye funkt͡sii. English
Description: [2016 edition]. | Providence, Rhode Island : American Mathematical Society : AMS Chelsea Publishing, 2016- | Originally published in Russian in 1958. | Originally published in English as 5 volume set: New York : Academic Press, 1964-[1968]. | Includes bibliographical references and index.
Identifiers: LCCN 2015040021 | ISBN 9781470426583 (v. 1 : alk. paper) | ISBN 9781470426590 (v. 2) | ISBN 9781470426613 (v. 3) | ISBN 9781470426620 (v. 4) | ISBN 9781470426637 (v. 5)
Subjects: LCSH: Theory of distributions (Functional analysis) | AMS: Functional analysis – Distributions, generalized functions, distribution spaces – Distributions, generalized functions, distribution spaces. msc
Classification: LCC QA331 .G373 2016 | DDC 515.7–dc23 LC record available at http://lccn.loc.gov/2015040021

♾ The paper used in this book is acid-free and falls within the guidelines established to ensure permanence and durability.
Visit the AMS home page at `http://www.ams.org/`

10 9 8 7 6 5 4 3 2 1 21 20 19 18 17 16

Translator's Note

According to the wish of Professor Gel'fand, this translation has been compared with the 1964 German translation,[1] and all improvements and omissions contained in the latter were taken over here. Certain minor corrections were made without being mentioned and a few notes were added (which are identified as translator's notes), especially in the last chapter, which is closely related to Chapter I in Volume 4 of this series.

No serious attempt has been made to coordinate the terminology with that used in previously published volumes, partly because the present translator does not entirely agree with it (e.g. the use of conjugate space for what is called here dual space, or function of bounded support, for what is more frequently called function of compact support). On the other hand, there are no radical departures from the notation and terminology of the authors—in particular, no attempt has been made to "modernize" it.[2]

The theory of partial differential equations, and of generalized eigenfunction expansions has made tremendous progress in the past few years. Not being a specialist in these fields the translator has made no attempt to update the literature on the subject (except for a few obvious references).

It was the express wish of Professor Gel'fand to refer the reader to the "excellent book of Hörmander" for some of the more recent developments.[3] This book is indeed the most valuable contribution to the literature on partial differential equations and should be read by any serious student of the subject.

Finally, I would like to thank Professor Gel'fand for supplying me with a copy of the German edition of this book and other literature which was useful in the translation.

October, 1967 MEINHARD E. MAYER

[1] "Verallgemeinerte Funktionen (Distributionen). Volume III—Einige Fragen zur Theorie der Differentialgleichungen." VEB-Deutscher Verlag der Wissenschaften, Berlin, 1964.

[2] As was done in the recently published French translation, *Math. Rev.* 1080 (1966), rev. Nr. 6001.

[3] L. Hörmander, "Linear Partial Differential Operators." Springer-Verlag, Berlin-Heidelberg-Göttingen and Academic Press, New York, 1963.

Preface to the Russian Edition

In the present volume, the third in the series "Generalized Functions," the apparatus of generalized functions is applied to the investigation of the following problems of the theory of partial differential equations: the problems of determining uniqueness and correctness classes for solutions of the Cauchy problem for systems with constant (or only time-dependent) coefficients and the problem of eigenfunction expansions for self-adjoint differential operators.

In subsequent volumes, the authors intend to discuss boundary value problems for elliptic equations and the Cauchy problem for equations with variable coefficients and for quasilinear equations, as well as problems related to complex extensions of all independent variables.

The authors use this occasion to thank the participants of the Seminar on Generalized Functions and Partial Differential Equations at Moscow State University, where various sections of this volume were repeatedly discussed. In particular, they are grateful to V. M. Borok, A. G. Kostyuchenko, Ya. I. Zhitomirskii and G. N. Zolotarev. The authors would also like to thank I. I. Shulishova for setting up detailed indexes for the first three volumes and to M. S. Agranovich, who has carefully edited the whole text and whose criticism has contributed considerable improvements.

Moscow, 1958

I. M. GEL'FAND
G. E. SHILOV

Contents

Chapter IV

CHAPTER I

SPACES OF TYPE W

This chapter contains an exposition of the theory of test function spaces of type W, which together with the spaces of type S (Volume 2, Chapter IV) will be used in Chapters II and III of the present volume for the study of Cauchy's problem. The results contained in the present chapter have been summarized without proofs in Appendix 2 to Chapter IV of Volume 2.

The spaces of type W are analogous to spaces of type S, corresponding to values $\alpha < 1$ and $\beta < 1$, but due to the use of arbitrary convex functions in place of powers, these spaces are capable of a more precise description of the peculiarities of growth (or decrease) at infinity.

In the same manner as for spaces of type S, for simplicity we shall first treat the case of one independent variable. The modifications which are necessitated by considering several independent variables are indicated in Section 4.

1. Definitions

1.1. The Spaces W_M

Let $\mu(\xi)$ $(0 \leqslant \xi < \infty)$ denote a continuous increasing function, such that $\mu(0) = 0$, $\mu(\infty) = \infty$. We define for $x \geqslant 0$

$$M(x) = \int_0^x \mu(\xi)\, d\xi. \tag{1}$$

The function $M(x)$ is an increasing convex continuous function, with $M(0) = 0$, $M(\infty) = \infty$. Since $\mu(\xi)$ increases with the increase of ξ, so does its average ordinate $x^{-1}M(x)$, so that for arbitrary positive x_1 and x_2, we have

$$\frac{1}{x_1} M(x_1) \leqslant \frac{1}{x_1 + x_2} M(x_1 + x_2),$$

$$\frac{1}{x_2} M(x_2) \leqslant \frac{1}{x_1 + x_2} M(x_1 + x_2).$$

Multiplying the first inequality by x_1, the second by x_2, and adding, we obtain the fundamental (convexity) inequality

$$M(x_1) + M(x_2) \leqslant M(x_1 + x_2). \tag{2}$$

In particular, for any $x \geqslant 0$

$$2M(x) \leqslant M(2x). \tag{3}$$

Further, we define the function $M(x)$ for negative x by means of the equality

$$M(-x) = M(x).$$

Note that since the derivative $\mu(x)$ of the function $M(x)$ is unbounded for $x \to +\infty$, the function $M(x)$ itself will *grow faster than any linear function* as $|x| \to \infty$.

We shall denote by W_M the set of all infinitely differentiable functions $\varphi(x)$ $(-\infty < x < \infty)$ satisfying the inequalities

$$|\varphi^{(q)}(x)| \leqslant C_q e^{-M(ax)} \tag{4}$$

with constants C_q and a which may depend on the function φ.

Since the function $M(x)$ increases faster than any linear function, the function $e^{-M(ax)}$ will decrease faster than any exponential function (i.e., a function of the form $e^{-a|x|}$); thus the test functions $\varphi(x)$ which belong to the space W_M, as well as all their derivatives, decrease at infinity faster than any exponential function.

It is obvious that W_M is a vector space (with the usual operations). We introduce for this space, the following definition of convergence: a sequence $\{\varphi_\nu(x)\}$ is said to *converge to zero* if the functions $\varphi_\nu(x)$ and all their derivatives converge to zero uniformly on any finite interval of the x-axis (such convergence is called *regular convergence*) and in addition the following inequalities hold:

$$|\varphi_\nu^{(q)}(x)| \leqslant C_q \, e^{-M(ax)}, \tag{5}$$

where the constants C_q and a do not depend on ν.

Let us show that the space W_M can be represented as a union of countably normed spaces.

We denote by $W_{M,a}$ the set of all functions from the space W_M which obey the inequalities

$$|\varphi^{(q)}(x)| \leqslant C_q \exp[-M(\bar{a}x)],$$

where the constant $\bar{a}$ can be selected arbitrarily, but smaller than a. In other words, the space $W_{M,a}$ consists of those functions $\varphi(x)$ which satisfy for any $\delta > 0$ the inequalities

$$|\varphi^{(q)}(x)| \leqslant C_{q\delta} \exp[-M[(a-\delta)x]] \qquad (q = 0, 1, 2,\ldots).$$

We define

$$M_p(x) = \exp\left(M\left[a\left(1-\frac{1}{p}\right)x\right]\right) \qquad (p = 2, 3,\ldots). \tag{6}$$

The functions $M_p(x)$ form an increasing sequence, $M_p(x) \leqslant M_{p+1}(x)$, and the functions $\varphi(x) \in W_{M,a}$ can be characterized as infinitely differentiable functions for which the norm

$$\|\varphi\|_p = \sup_{\substack{x \\ |q| \leqslant p}} M_p(x)\,|\varphi^{(q)}(x)| \tag{7}$$

is finite for arbitrary p. This shows that the space $W_{M,a}$ coincides with the space $K\{M_p\}$ defined in Volume 2, Chapter II, Section 1, with a fixed sequence of weight functions (6). Therefore, all the results referring to the spaces $K\{M_p\}$ may be applied to the space $W_{M,a}$. It is thus a complete countably normed space with the norms (7). We show that it is a perfect space. The condition (p), which is sufficient for the space $K\{M_p\}$ to be perfect (Volume 2, Chapter II, Section 2), consists in the existence for any p of a number $p' > p$, such that

$$\lim_{|x|\to\infty} \frac{M_p(x)}{M_{p'}(x)} = 0.$$

In our case, due to the convexity inequality, we have for any $p' > p$

$$M\left[a\left(1-\frac{1}{p}\right)x\right] + M\left[a\left(\frac{1}{p}-\frac{1}{p'}\right)x\right] \leqslant M\left[a\left(1-\frac{1}{p'}\right)x\right],$$

and consequently

$$\begin{aligned}\frac{M_p(x)}{M_{p'}(x)} &= \exp\left(M\left[\left(1-\frac{1}{p}\right)ax\right] - M\left[\left(1-\frac{1}{p'}\right)ax\right]\right)\\ &\leqslant \exp\left(-M\left[\left(\frac{1}{p}-\frac{1}{p'}\right)ax\right]\right) \to 0,\end{aligned}$$

as required for the proof.

According to the results of Volume 2, Chapter II, Section 2, a sequence $\varphi_\nu(x) \in W_{M,a}$ converges to zero if and only if the sequence $\varphi_\nu(x)$ is regularly convergent (i.e., the functions $\varphi_\nu^{(q)}(x)$, for any q, converge uniformly to zero on any interval $|x| \leqslant x_0 < \infty$) and the norms $\|\varphi_\nu\|_p$ are bounded for any p.

The union of the spaces $W_{M,a}$ with all indices $a = 1, \frac{1}{2}, \dots$ obviously coincides with the space W_M. The convergence to zero in the space W_M, as described above, is the convergence to zero in one of the spaces $W_{M,a}$, and thus coincides with the concept of convergence defined in W_M considered as a union of countably normed spaces.

We now define bounded sets in the space W_M. According to the general definition of bounded sets in a union of countably normed spaces, set $A \subset W_M$ is said to be *bounded*, if it is entirely contained within one of the $W_{M,a}$ and is bounded in this space. In other words, the set $A \subset W_M$ is bounded if all functions $\varphi(x) \in A$ satisfy the inequalities (4) with the same constants C_q and a. In particular, a sequence $\varphi_\nu(x) \in W_M$ converges to zero if (1) it converges to zero regularly, (2) it is bounded.

Example 1. Let $M(x) = x^{1/\alpha}$ $(x > 0)$, with $\alpha < 1$; then $\mu(\xi) = (1/\alpha)\, \xi^{(1/\alpha)-1}$. The corresponding space W_M consists of infinitely differentiable functions $\varphi(x)$, satisfying the inequalities

$$|\varphi^{(q)}(x)| \leqslant C_q e^{-a|x|^{1/\alpha}}$$

for certain C_q and a which depend on φ. Obviously, this space coincides with the space S_α (Volume 2, Chapter IV, Section 1).

Example 2. Let $\mu(\xi) = \ln(\xi + 1)(\xi \geqslant 0)$; then, for $x \geqslant 0$

$$M(x) = \int_0^x \ln(\xi + 1)\, d\xi = (x + 1)\ln(x + 1) - x.$$

According to the definition, the space W_M consists of the infinitely differentiable functions $\varphi(x)$ which satisfy the inequalities

$$|\varphi^{(q)}(x)| \leqslant C_q \exp(-a[(|x| + 1)\ln(|x| + 1) - |x|]).$$

In this case the functions $\varphi(x)$ admit a simpler description, which can be obtained by means of the following considerations.

Formally one could have constructed a space W_M starting from any nonnegative continuous function $M(x)$ (without taking into account whether this function has the special form (1); later we shall make use of this special form), by means of the definition (4). In this case, one may obtain the same space for different functions $M_1(x)$ and $M_2(x)$, $W_{M_1} \equiv W_{M_2}$. We indicate a simple sufficient condition for this equality to hold. Assume that the functions $M_1(x)$ and $M_2(x)$ satisfy for sufficiently large $x \geqslant 0$ the inequality

$$M_1(\gamma_1 x) \leqslant M_2(\gamma_2 x) \tag{8}$$

with some positive constants γ_1 and γ_2. Then we can assert that the inclusion

$$W_{M_1} \supset W_{M_2}$$

holds. Indeed, Eq. (8) can be replaced by an inequality, valid for all values of $x \geqslant 0$, by adding a suitable constant

$$M_1(\gamma_1 x) \leqslant M_2(\gamma_2 x) + \gamma_3 .$$

Hence, if $\varphi(x) \in W_{M_2}$, we have

$$|\varphi^{(q)}(x)| \leqslant C_q \exp[-M_2(ax)] \leqslant C_q' \exp[M_1(a'x)], \tag{9}$$

with $a' = a(\gamma_1/\gamma_2)$, $C_q' = C_q e^{\gamma_3}$; thus $\varphi \in W_{M_1}$.

Moreover, the inequality (9) shows that if the sequence $\varphi_\nu(x)$ converges to zero in the sense of W_{M_2}, it does so also in the sense of W_{M_1}, since one can choose the constants a' and C_q' in the inequalities (9) for the functions $\varphi_\nu(x)$ together with the constants a and C_q independently of ν.

Further, if the functions $M_1(x)$ and $M_2(x)$ are such that for sufficiently large $x \geqslant 0$

$$M_1(\gamma_1 x) \leqslant M_2(\gamma_2 x) \leqslant M_1(\gamma_1' x), \tag{10}$$

then both inclusions $W_{M_1} \supset W_{M_2}$ and $W_{M_2} \supset W_{M_1}$ hold, and thus $W_{M_1} \equiv W_{M_2}$, as sets; it is also obvious that the convergence in W_{M_1} coincides with the convergence in W_{M_2}. Two functions $M_1(x)$ and $M_2(x)$ satisfying the inequality (10) will be called *equivalent*; we have seen that equivalent functions define the same space.

The function $(x + 1)\ln(x + 1) - x$, which appears in Example 2, is equivalent to the function $x \ln x$ (which does not satisfy the definition (1)); consequently, the corresponding test function space is also defined by means of the inequalities

$$|\varphi^{(q)}(x)| \leqslant C_q \exp(-a\,|x|\ln|x|)$$

and the corresponding definition of convergence.

1.2. The Spaces W^{Ω}

Let $\omega(\eta)$ $(0 \leqslant \eta < \infty)$ denote an increasing continuous function with $\omega(0) = 0$, $\omega(\infty) = \infty$; for $y \geqslant 0$ we define

$$\Omega(y) = \int_0^y \omega(\eta)\, d\eta. \tag{1}$$

The properties of the function $\Omega(y)$ are entirely analogous to those of the function $M(x)$, introduced in Section 1.1; in particular, the convexity inequalities hold:

$$\Omega(y_1) + \Omega(y_2) \leqslant \Omega(y_1 + y_2), \tag{2}$$

$$2\Omega(y) \leqslant \Omega(2y) \tag{3}$$

We further define

$$\Omega(-y) = \Omega(y).$$

We shall denote by W^{Ω} the set of all entire analytic functions $\varphi(z)$ $(z = x + iy)$, which satisfy the inequalities

$$| z^k\varphi(z) | \leqslant C_k e^{\Omega(by)} \tag{4}$$

where the constants C_k and b depend on the function φ.

It is obvious that W^{Ω} is a vector space with the usual definitions of the operations (over the field of complex numbers). We introduce for this vector space the following definition of convergence: a sequence $\varphi_\nu(z) \in W^{\Omega}$ is said to *converge to zero* if the functions $\varphi_\nu(z)$ converge uniformly to zero in any bounded domain of the z-plane (this will be called regular convergence) and in addition satisfy the inequalities

$$| z^k\varphi_\nu(z) | \leqslant C_k e^{\Omega(by)},$$

where the constants C_k and b do not depend on the index ν.

The space W^{Ω} can be represented as a union of countably normed spaces. Indeed, let us denote by $W^{\Omega,b}$ the set of those functions in W^{Ω} which satisfy the inequalities

$$| z_k\varphi(z) | \leqslant C_k \exp[\Omega(\bar{b}y)],$$

where $\bar{b}$ can be any constant larger than b. In other words, the set $W^{\Omega,b}$ consists of those entire functions which for any $\rho > 0$ satisfy the inequalities

$$|z_k\varphi(z) | \leqslant C_{k\rho} \exp[\Omega[(b + \rho)y]).$$

In the space $W^{\Omega,b}$ we define the norms

$$\| \varphi \|_{kp} = \sup_z | z^k \varphi(z) | \exp(-\Omega[(b + \rho)y]). \quad (5)$$

Let us show, that with the norms (5) the space $W^{\Omega,b}$ becomes a complete, perfect, countably normed space.

1. *The norms* $\| \varphi \|_{k\rho}$ *agree with each other.* Indeed, if the sequence $\varphi_\nu \in W^{\Omega,b}$ is fundamental with respect to two norms $\| \varphi_\nu \|_{k\rho}$ and $\| \varphi_\nu \|_{k_1\rho_1}$ (i.e., satisfies in each of these norms the Cauchy condition) and converges to zero with respect to one of the two norms, then, in any case, the functions $\varphi_\nu(z)$ converge to zero at each point; it follows that the limit of the sequence is zero also with respect to the second norm.

2. *The space* $W^{\Omega,b}$ *is complete.* The completion of the set $W^{\Omega,b}$ with respect to one of the norms $\| \varphi \|_{k\rho}$ consists of all entire analytic functions with finite values of $\| \varphi \|_{k\rho}$. The intersection of all these completions with respect to the indices k and ρ consists of those entire analytic functions, for which $\| \varphi \|_{k\rho}$ is meaningful for any k and ρ, i.e., coincides with the space $W^{\Omega,b}$. According to the theorem in Volume 2, Chapter I, Section 3.2, this fact guarantees the completeness of the space.

3. *The space* $W^{\Omega,b}$ *is perfect.* The proof of an analogous assertion for the space $\Phi = K\{M_p\}$ (given in Volume 2, Chapter II, Section 2) was based on the condition that any sequence of test functions $\{\varphi_\nu(x)\}$ which is bounded with respect to all norms of the space and which converges regularly (i.e., converges uniformly on any finite interval, together with all the derivatives) also converges in the topology of the space Φ.

In the present case it is easy to verify that this assumption holds. It follows that the result also holds, i.e., the space $W^{\Omega,b}$ is perfect.

Obviously, the union of all countably normed spaces $W^{\Omega,b}$ with $b = 1, 2,...$ coincides with the space W^{Ω}. The convergence to zero in the space W^{Ω}, as described above, is the same as convergence to zero in one of the spaces $W^{\Omega,b}$ and consequently coincides with the concept of convergence to zero which is defined in W^{Ω} by considering it as the union of the spaces $W^{\Omega,b}$.

We define bounded sets in the space W^{Ω}: according to the general definition of a bounded set in a union of countably normed spaces

(Volume 2, Chapter I, Section 8), a set $A \subset W^{\Omega}$ is said to be *bounded* if it is entirely contained in one of the $W^{\Omega,b}$ and is bounded in that latter space. In other words, the set $A \subset W^{\Omega}$ is bounded if for all functions $\varphi(z) \in A$ the inequalities (4) are fulfilled with the same C_k and b. Thus, a sequence $\varphi_\nu \in W^{\Omega}$ converges to zero if: (1) it converges regularly to zero, (2) it is bounded.

In the same manner as in Section 1.1 one can show that for equivalent functions $\Omega_1(y)$ and $\Omega_2(y)$ the spaces W^{Ω_1} and W^{Ω_2} coincide as sets as well as in topology.

We shall have occasion to use later the following particular examples of W^{Ω} spaces.

Example 1. Let $\omega(\eta) = (1-\beta)^{-1}\eta^{1/(1-\beta)-1}$ for $\eta > 0$, $\Omega(y) = y^{-1/(1-\beta)}$ ($\beta < 1$). The space W^{Ω} consists of all entire analytic functions satisfying the inequalities

$$| z^k\varphi(z) | \leqslant C_k \exp(b \mid y \mid)^{1/(1-\beta)},$$

and therefore coincides with the space S^{β} (Volume 2, Chapter IV, Section 2).

Example 2. Let $\omega(\eta) = e^{\eta} - 1$ for $\eta > 0$, $\Omega(y) = \int_0^y (e^{\eta} - 1)\, d\eta = e^y - y - 1$. This function $\Omega(y)$ is equivalent to the function $\Omega_1(y) = e^y$. Therefore the space W^{Ω} can in this case be defined by means of the inequality

$$| z^k\varphi(z) | \leqslant C_k \exp(e^{b|y|})$$

where the constant C_k and b depend on the function φ, together with the appropriate convergence.

1.3. The Spaces $W_M{}^{\Omega}$

Let $\mu(\xi)$ and $\omega(\eta)$ ($0 \leqslant \xi, \eta < \infty$) denote a pair of increasing continuous functions; define for $x \geqslant 0, y \geqslant 0$:

$$M(x) = \int_0^x \mu(\xi)\, d\xi, \qquad \Omega(y) = \int_0^y \omega(\eta)\, d\eta$$

and for $x < 0$, $y < 0$

$$M(x) = M(-x), \qquad \Omega(y) = \Omega(-y).$$

The functions $M(x)$ and $\Omega(y)$ are the same as those introduced in Sections 1.1 and 1.2.

We denote by $W_M{}^\Omega$ the set of all entire analytic functions $\varphi(z)$ ($z = x + iy$) which satisfy the inequalities

$$|\varphi(x + iy)| \leqslant C \exp[-M(ax) + \Omega(by)] \tag{1}$$

where the constants a, b, and C depend on the function $\varphi(z)$.

Obviously $W_M{}^\Omega$ is a vector space with the usual operations. We define for this space the following concept of convergence: a sequence $\varphi_\nu(z) \in W_M{}^\Omega$ is said to ***converge to zero*** if the functions $\varphi_\nu(z)$ converge uniformly to zero in any bounded domain in the z-plane and in addition the inequalities

$$|\varphi_\nu(z)| \leqslant C \exp[-M(ax) + \Omega(by)]$$

hold, with constants C, a, b which do not depend on the index ν.

The space $W_M{}^\Omega$ can also be represented as a union of countably normed spaces. We denote by $W_{M,a}^{\Omega,b}$ the set of those functions of the space $W_M{}^\Omega$ which satisfy the inequalities

$$|\varphi(x + iy)| \leqslant C \exp[-M(\bar{a}x) + \Omega(\bar{b}y)],$$

where $\bar{a}$ is any constant smaller than a, and $\bar{b}$ is any constant larger than b. For $\varphi \in W_{M,a}^{\Omega,b}$ we define

$$\|\varphi\|_{\delta\rho} = \sup_z |\varphi(z)| \exp(M[(a - \delta)x] - \Omega[(b + \rho)y]).$$

These norms evidently satisfy all axioms required of a system of norms. The proof of the fact that with this system of norms $W_{M,a}^{\Omega,b}$ becomes a perfect countably normed space is identical to the proof carried out for the case of the space $W^{\Omega,b}$.

The union of all spaces $W_{M,a}^{\Omega,b}$ with respect to all $a = 1, \frac{1}{2},\ldots$ and $b = 1, 2,\ldots$ obviously coincides with the space $W_M{}^\Omega$, and the convergence in the latter space is defined in the same manner as in a union of countably normed spaces. In the same manner as in the spaces $W_{M,a}$ and $W^{\Omega,b}$, a sequence $\varphi_\nu(x) \in W_M{}^\Omega$ converges to zero if and only if: (i) it converges regularly to zero, and (ii) it is bounded. A set $A \subset W_M{}^\Omega$ is said to be ***bounded*** if it is entirely contained in some $W_{M,a}^{\Omega,b}$ and is bounded in this space; in other words, the set $A \subset W_M{}^\Omega$ is bounded if all functions $\varphi(z) \in A$ satisfy the inequalities (1) with the same values of a, b, C.

Equivalent functions $M_1(x)$ and $M_2(x)$, $\Omega_1(y)$ and $\Omega_2(y)$ (cf. Section 1.1) define the same space: $W_{M_1}^{\Omega_1} = W_{M_2}^{\Omega_2}$, both as sets and in topology.

Examples. Combining the functions which were introduced earlier: $M_1(x) = x^{1/\alpha}$, $M_2(x) = x \ln x$, $\Omega_1(y) = y^{1/(1-\beta)}$, $\Omega_2(y) = e^y$ ($\alpha < 1$, $\beta < 1$), we obtain the four spaces:

$$W_{M_1}^{\Omega_1}, \quad W_{M_1}^{\Omega_2}, \quad W_{M_2}^{\Omega_1}, \quad W_{M_2}^{\Omega_2}.$$

The space $W_{M_1}^{\Omega_1}$ consists of the entire functions $\varphi(x + iy)$ satisfying the inequalities

$$| \varphi(x + y) | \leqslant C \exp[-a | x |^{1/\alpha} + b | y |^{1/(1-\beta)}],$$

and consequently coincides with the space $S_\alpha^{\ \beta}$ (Volume 2, Chapter IV, Section 2). In the three remaining cases new test function spaces are obtained, with the functions satisfying the inequalities

$$W_{M_1}^{\Omega_2}: \quad | \varphi(x + iy) | \leqslant C \exp[-a | x |^{1/\alpha} + e^{b|y|}];$$

$$W_{M_2}^{\Omega_1}: \quad | \varphi(x + iy) | \leqslant C \exp[-a | x \ln x | + b | y |^{1/(1-\beta)}],$$

$$W_{M_2}^{\Omega_2}: \quad | \varphi(x + iy) | \leqslant C \exp[-a | x \ln x | + e^{b|y|}].$$

In what follows we shall denote by $W_{r,a}^{s,q}$ the space $W_{M,a}^{\Omega,b}$ with $M(x) = x^r$, $\Omega(y) = y^s$ ($r > 1, s > 1$). A similar sense should be attributed to the notation $W_r^{\ s}$.

1.4. The Problem of Nontriviality of the Spaces W_M

In the same manner as in the spaces $S_\alpha^{\ \beta}$, some spaces $W_M^{\ \Omega}$ may turn out to be trivial (i.e., consisting only of the single function $\varphi(z) \equiv 0$). For instance, this will always be so, if

$$\lim_{x \to \infty} [\Omega(bx) - M(ax)] = -\infty \tag{1}$$

for arbitrary a and b. Indeed, let (1) be true and $\varphi \in W_M^{\ \Omega}$. This means that for some a and b

$$| \varphi(z) | = | \varphi(x + iy) | \leqslant C \exp[-M(ax) + \Omega(by)]$$

Then, also

$$| \varphi(iz) | = | \varphi(ix - y) | \leqslant C \exp[-M(ay) + \Omega(bx)];$$

whence

$$| \varphi(z)\varphi(iz) | \leqslant C^2 \exp[-M(ax) + \Omega(bx)] \exp[-M(ay) + \Omega(by)].$$

According to condition (1), the function $\varphi(z)\,\varphi(iz)$ is bounded for sufficiently large $|z|$, and hence, by Liouville's theorem, it is constant. Since, in addition, (1) implies that $\lim_{x\to\infty}\varphi(x)\,\varphi(ix) = 0$, it follows that $\varphi(z)\,\varphi(iz) \equiv 0$, and hence $\varphi(z) \equiv 0$.

A class of nontrivial spaces of type W will be indicated below.

A continuous function $l(x)$ $(x > 0)$ will be called a *slow function*, if for any $\epsilon > 0$ and sufficiently large $x > x_0 = x_0(\epsilon)$ the inequality

$$C_\epsilon' x^{-\epsilon} < l(x) < C_\epsilon x^\epsilon. \tag{2}$$

is verified.

B. Ya. Levin has proved the following theorem, which is one of the results of his theory of the generalized growth indicatrix for entire functions of finite order.

For arbitrary $p > 0$ and any slow function $l(x)$, there exists an entire analytic function $\varphi(z) \not\equiv 0$, for which

$$|\varphi(x+iy)| \leqslant C\exp[-l(x)\,|x|^p + \gamma l(y)\,|y|^p]$$

with some constants C and γ.

This theorem implies the nontriviality of the space $W_M{}^\Omega$ with $M = \Omega = l(x)\,x^p$, where $l(x)$ is a slow function.

Obviously, at the same time the nontriviality is established for any space $W_M^{\Omega_1}$, with $M = l(x)\,x^p$, $\Omega_1(x) \geqslant l(x)\,x^p$, since the latter space contains the notrivial space $W_M{}^\Omega$.

1.5. On the "Richness" in Functions of the Spaces W_M

Let us assume that a given space $W_M{}^\Omega$ is nontrivial. Therefore, for some $a > 0$, $b > 0$, there certainly exist nontrivial spaces $W_{M,a}^{\Omega,b}$; we shall call the corresponding pairs of numbers (a, b) "admissible." Since for $a \leqslant a_0$, $b \geqslant b_0$ we have $-M(a_0x) \leqslant -M(ax)$, $\Omega(b_0y) \leqslant \Omega(by)$, the region of admissible values of a and b will contain together with any pair a_0, b_0, all pairs with $a \leqslant a_0$, $b \geqslant b_0$. Replacing the function $\varphi(z) \in W_{M,a}^{\Omega,b}$ by $\varphi(\lambda z)$ with positive λ, one can see that, together with any admissible pair (a, b), the pair $(\lambda a, \lambda b)$ will also be admissible. Thus the region of admissible pairs (a, b), together with any pair (a_0, b_0), contains all pairs which satisfy the relation $b/a \geqslant b_0/a_0$. It follows that the total region of admissible pairs (a, b), in the quadrant $a \geqslant 0$, $b \geqslant 0$ of the (a, b)-plane, represents a certain angle defined by the inequality $\tan(b/a) \geqslant \gamma$ (or by the inequality $\tan(b/a) > \gamma$), where γ depends on the functions $M(x)$ and $\Omega(y)$.

All nontrivial spaces of type W are sufficiently "rich" in functions, in the sense indicated in Volume 2, Chapter IV, Section 8; namely, if for some locally integrable function $f(x)$ and any test function $\varphi(x)$

$$\int_{-\infty}^{\infty} f(x)\varphi(x)\,dx = 0,$$

then $f(x) \equiv 0$ almost everywhere.

The proof can be given following the scheme of reasoning used for the proof of the same property of spaces of type S (Volume 2, Chapter IV, Section 8).

On the basis of the result of Volume 2, Chapter IV, Section 8.8, it follows that any nontrivial space of type W is densely embedded in any normed function space $E \supset W$, with the norm defined as

$$\|\varphi\| = \int_{-\infty}^{\infty} M(x)\,|\,\varphi(x)\,|\,dx.$$

2. Bounded Operators in Spaces of Type W

In this section we show that in the spaces of type W which were introduced in Section 1, the operations of differentiation and multiplication by the variable x, as well as multiplication by some entire analytic functions, are defined and continuous.

2.1. Operations in the Space W_M

1. THE OPERATION OF DIFFERENTIATION

Let $\varphi(x) \in W_M$, so that

$$|\,\varphi^{(q)}(x)\,| \leqslant C_q e^{-M(ax)}.$$

Then the function $\varphi_1(x) = \varphi'(x)$ will satisfy the inequalities

$$|\,\varphi_1^{(q)}(x)\,| = |\,\varphi^{(q+1)}(x)\,| \leqslant C_{q+1} e^{-M(ax)},$$

i.e., the function $\varphi_1(x)$ also belongs to the space W_M. Obviously any bounded set in the space W_M is transformed by differentiation into another bounded set. Thus the operation of differentiation is bounded in W_M and consequently is a continuous operation.

2. The Operation of Multiplication by x

Let $\varphi(x) \in W_M$, so that $|\varphi^{(q)}(x)| \leqslant C_q e^{-M(ax)}$. Then the function $\varphi_1(x) = x\varphi(x)$ satisfies the inequalities

$$| \varphi_1^{(q)}(x) | = | [x\varphi(x)]^{(q)} | \leqslant | x\varphi^{(q)}(x) | + q | \varphi^{(q-1)}(x) |$$

$$\leqslant | x | C_q e^{-M(ax)} + qC_{q-21} e^{-M(ax)}.$$

For any positive δ, the inequality

$$| x | e^{-M(ax)} \leqslant C_\delta e^{-M[(a-\delta)x]} \tag{1}$$

holds; indeed, according to the convexity condition (Eq.(2) in Section 1.1)

$$| x | e^{-M(ax)+M[(a-\delta)x]} \leqslant | x | e^{-M(\delta x)},$$

and thus the quantity which has been obtained is bounded, due to the exponential decrease of $e^{-M(\delta x)}$.

Making use of the inequality (1) we obtain

$$| \varphi_1^{(q)}(x) | \leqslant C_q C_\delta e^{-M[(a-\delta)x]} + qC_{q-1} e^{-M(ax)} \leqslant C_q' e^{-M[(a-\delta)x]},$$

where $C_q' = C_q C_\delta + qC_{q-1}$. It follows that the function $\varphi_1(x)$ belongs to the space W_M.

It is also obvious that any bounded set in the space W_M is taken into a bounded set by multiplication with x. Thus the operation of multiplication by x is bounded on the space W_M and consequently is a continuous operation.

2.2. Operations in the Space W^Ω

1. The Operation of Differentiation

Let $\varphi(z) \in W^\Omega$, so that

$$| z^k \varphi(z) | \leqslant C_k e^{\Omega(by)}$$

Assume that $\varphi_1(z) = \varphi'(z)$. Then

$$| z^k \varphi_1(z) | = | [z^k \varphi(z)]' - kz^{k-1}\varphi(z) |$$

$$\leqslant | [z^k \varphi(z)]' | + k | z^{k-1}\varphi(z) |;$$

further, applying Cauchy's differentiation formula, with the integral taken over a circle Γ of radius r centered in the point $z = x + iy$, $y \geqslant 0$,

$$[z^k\varphi(z)]' = \frac{1}{2\pi i}\int_\Gamma \frac{\xi^k\varphi(\xi)\,d\xi}{(\xi - z)^2};$$

it follows that

$$|\,[z^k\varphi(z)]'\,| \leqslant (1/r)\max_\Gamma |\,\xi^k\varphi(\xi)\,| \leqslant (1/r)C_k e^{\Omega(b(y+r))}.$$

Since the function $\Omega(y)$ increases monotonically, and for sufficiently large y,

$$b(y + r) \leqslant (b + r)y,$$

we will also have for the same values of y

$$\Omega[b(y + r)] \leqslant \Omega[(b + r)y].$$

The following inequality certainly holds for all values of y:

$$\Omega[b(y + r)] \leqslant \Omega[(b + r)y] + C_r\,.$$

Thus, for $y \geqslant 0$

$$|\,[z^k\varphi(z)]'\,| \leqslant C_{kr}e^{\Omega[(b+r)y]},$$

where

$$C_{kr} = (1/r)C_k e^{C_r}\,.$$

Further,

$$k\,|\,z^{k-1}\varphi(z)\,| \leqslant kC_{k-1}e^{\Omega(by)} \leqslant kC_{k-1}e^{\Omega[(b+r)y]};$$

as a result of the preceding inequalities we obtain:

$$|\,z^k\varphi_1(z)\,| \leqslant C_{kr}e^{\Omega[(b+r)y]} + kC_{k-1}e^{\Omega[(b+r)y]} \leqslant C'_{kr}e^{\Omega[(b+r)y]};$$

which means that $\varphi_1(z)$ belongs to the space W^Ω. A similar evaluation can be carried out also for $y < 0$. Obviously, the operation of differentiation carries a bounded set of the space W^Ω into another bounded set. Thus differentiation is a bounded, and hence continuous operation on the space W^Ω.

2. The Operation of Multiplication by z

Let $\varphi(z) \in W^\Omega$, so that

$$|\,z^k\varphi(z)\,| \leqslant C_k e^{\Omega(by)}.$$

Putting $\varphi_1(z) = z\varphi(z)$, we obtain

$$| z^k\varphi_1(z) | = | z^{k+1}\varphi(z) | \leqslant C_{k+1}e^{\Omega(by)},$$

i.e., $\varphi_1(z)$ also belongs to the space W^Ω. It is obvious that multiplication by z takes any bounded set in W^Ω into another bounded set. Thus the operation of multiplication by z is bounded, and hence continuous in W^Ω.

2.3. Operations in the Space $W_M{}^\Omega$

1. The Operation of Differentiation

Let $\varphi(z) \in W_M{}^\Omega$, so that

$$| \varphi(z) | \leqslant Ce^{-M(ax)+\Omega(by)}.$$

We define $\varphi_1(z) = \varphi'(z)$; making use of the Cauchy formula with the same circle Γ of radius r ($< a$) as before, we find

$$| \varphi_1(z) | \leqslant (1/r) \max_\Gamma | \varphi(\xi) | \leqslant (C/r) \exp\{M[(x - r)a] + \Omega[b(y + r)]\}$$

(for simplicity we have assumed that $x \geqslant r$, $y \geqslant 0$). In the same manner as in Section 2.2.1 we have the inequality

$$\Omega[b(y + r)] \leqslant C_r' + \Omega[(b + r)y]$$

and similarly

$$M[a(x - r)] \geqslant C_r'' + M[(a - r)x];$$

therefore we obtain

$$| \varphi_1(z) | \leqslant C_r \exp\{-M[(a - r)x] + \Omega[(b + r)y]\}.$$

Consequently, $\varphi_1(z)$ belongs to the space $W_M{}^\Omega$. As before, differentiation is a bounded, and hence continuous operation in the space $W_M{}^\Omega$.

2. The Operation of Multiplication by z

Let $\varphi(z) \in W_M{}^\Omega$, so that

$$| \varphi(z) | \leqslant C \exp[-M(ax) + \Omega(by)].$$

We define $\varphi_1(z) = z\varphi(z)$. Then

$$\begin{aligned} | \varphi_1(z) | = | z\varphi(z) | &\leqslant C | z | \exp[M(ax) + \Omega(by)] \\ &\leqslant C | x | \exp[-M(ax)] + C | y | \exp[\Omega(by)] \end{aligned}$$

In the same manner as in Section 2.1.1 we derive the inequality

$$|\varphi_1(z)| \leqslant C_r \exp[-M[(a-r)x] + \Omega[(b+r)y]].$$

and similarly

$$|y| \exp\{\Omega(by)\} \leqslant C_\rho \exp\{\Omega[(b+\rho)y\};$$

therefore

$$|\varphi_1(z)| \leqslant C_{\delta\rho} \exp\{-\Omega[(a-\delta)x] + \Omega[(b+\rho)y]\};$$

which implies that $\varphi_1(z) \in W_M{}^\Omega$. Obviously, multiplication by z is a bounded, and hence continuous operation in the space $W_M{}^\Omega$.

In fact, the reasoning we used above shows that the operations of differentiation and multiplication by the independent variable are also defined and bounded (and hence continuous) in the "finer" spaces $W_{M,a}$, $W^{\Omega,b}$, and $W_{M,a}^{\Omega,b}$.

2.4. The Operation of Multiplication by Entire Analytic Functions

Theorem 1. *Let the entire analytic function $f(z)$ satisfy the inequality*

$$|f(z)| \leqslant Ce^{\Omega_0 b(y)}(1 + |x|^h). \tag{1}$$

Then the function $f(z)$ is a multiplier in the space W^Ω and at the same time takes the space $W^{\Omega,b}$ into the space $W^{\Omega,b+b_0}$.

Proof. By definition $\varphi(z) \in W^{\Omega,b}$, and according to the inequality (1) we have for any $\rho > 0$

$$|z^k f(z)\varphi(z)| \leqslant C_k \exp\{\Omega[(b+\rho)y]\}\, C \exp[\Omega(b_0 y)](1 + |x|^h). \tag{2}$$

Making use of the convexity condition, Eq. (2), Section 1.1, we obtain

$$|z^k f(z)\varphi(z)| \leqslant C_k'(1 + |x|^h) \exp[\Omega[(b + b_0 + \rho)y]]. \tag{3}$$

This inequality is valid for any $k = 0, 1, 2,\ldots$; replacing[1] k by $k + h$, we obtain the inequality

$$|z^k f(z)\varphi(z)| \leqslant C_k'' \frac{(1 + |x|^h)}{|x|^h} \exp[\Omega[(b + b_0 + \rho)y]]. \tag{4}$$

[1] If h is not an integer one may choose in this calculation the closest larger integer.

Combining (3) and (4) we have

$$| z^k f(z)\varphi(z) | \leqslant \exp[\Omega[(b + b_0 + \rho)y]] \min \left\{ C_k'(1 + | x |^h), C_k'' \frac{1 + | x |^h}{| x |^h} \right\}$$

$$\leqslant C_k''' \exp[\Omega[(b + b_0 + \rho)y]],$$

i.e., the product $f(z)\,\varphi(z)$ belongs to the space $W^{\Omega,b+b_0}$. Making use of the relation between the constants ($C_k''' \leqslant 2(C_k' + C_k'')$) it can be seen that the operator of multiplication by the function $f(z)$ is bounded in the space $W^{\Omega,b}$.

Theorem 2. *Let $M(x)$ and $\Omega(y)$ be the functions defining the space $W_M{}^\Omega$. Then the entire analytic function $f(z)$ satisfying the inequality*

$$| f(z) | \leqslant C \exp[M(a_0 x) + \Omega(b_0 y)] \tag{5}$$

defines a bounded multiplication operator in the space $W_{M,a}^{\Omega,b}$ for $a > a_0$ and takes the space $W_{M,b}^{\Omega,b}$ into the space $W_{m,a-a_0}^{\Omega,b+b_0}$

Proof. According to the definition of the function $\varphi(z) \in W_{M,a}^{\Omega,a}$ and the inequality (5), we have, for aıbitrary δ and ρ

$$| f(z)\varphi(z) | \leqslant CC_{\delta\rho} \exp[M(a_0 x) - M[(a - \delta)x] \exp(\Omega(b_0 y) + \Omega[(b + \rho)y])].$$

Making use of the convexity inequality (2), Section 1.1, we have

$$M(a_0 x) - M[(a - \delta)x] \leqslant - M[(a - a_0 - \delta)x],$$

$$\Omega(b_0 y) + \Omega[(b + \rho)y] \leqslant \Omega[(b + b_0 + \rho)y].$$

Thus we obtain

$$| f(z)\varphi(z) | \leqslant C' \exp[-M[(a - a_0 - \delta)x] + \Omega[(b + b_0 + \rho)y]]. \tag{6}$$

Hence, the product $f(z)\,\varphi(z)$ belongs to the space $W_{M,a-a_0}^{\Omega,b+b_0}$. The relations among the constants show that the operator of multiplication is bounded in the space $W_{M,a}^{\Omega,b}$.

The following remark completes the theorems we have proved:

Let the entire analytic function $f(z)$ satisfy, for any $\epsilon > 0$, an inequality of the form

$$| f(z) | \leqslant C_\epsilon e^{\Omega(\epsilon y)} \tag{7}$$

or of the form

$$| f(z) | \leqslant C_\epsilon e^{M(\epsilon x) + \Omega(\epsilon y)}. \tag{8}$$

Then the function $f(z)$ is a multiplier in any space $W^{\Omega,b}$ or $W_{M,a}^{\Omega,b}$, respectively.

The proof is a repetition of the proofs of Theorems 1 or 2, respectively, with a_0 and b_0 replaced by ϵ and making use of the conclusion that in this case the inequality (6) implies $f\varphi \in W_{M,a}^{\Omega,b}$.

Example. For any real σ the function $f(z) = e^{i\sigma z}$ is a multiplier in any of the spaces $W^{\Omega,b}$, $W_{M,a}^{\Omega,b}$.

Indeed, the inequality

$$| e^{i\sigma z} | \leqslant e^{|\sigma||y|}.$$

holds. But for any convex functions $M(x)$ and $\Omega(y)$ and arbitrary $\epsilon > 0$ one can write the inequalities

$$| \sigma | | y | \leqslant \Omega(\epsilon y) + C_1 ,$$
$$| \sigma | | y | \leqslant M(\epsilon x) + \Omega(\epsilon y) + C_1 ,$$

which, for $f(z) = e^{i\sigma z}$ imply the inequalities (7) and (8); thus $f(z)$ is indeed a multiplier in the spaces $W^{\Omega,b}$ or $W_{M,a}^{\Omega,b}$, as stated.

3. Fourier Transforms

We recall that a test function space Ψ is called Fourier-dual[2] with respect to a given test function space Φ, if Ψ consists of those functions $\psi(\sigma)$ which are Fourier transforms of the functions $\varphi \in \Phi$

$$\psi(\sigma) = \int_{-\infty}^{\infty} \varphi(x) e^{ix\sigma} \, dx.$$

In this section we shall indicate the spaces which are Fourier-dual to the spaces $W_{M,a}$, $W^{\Omega,b}$, and $W_{M,a}^{\Omega,b}$. As in the previous volumes, we shall denote the Fourier-dual of a space Φ by $\tilde{\Phi}$.

3.1. Dual Functions

We first introduce the important concept of functions which are *dual in the sense of Young*. Let the functions $M(x)$ and $\Omega(y)$ be defined by Eqs. (1) of Section 1.1 and (1) of Section 2.1, respectively. If the functions

[2] The Russian term "dvoistvennyi" has been translated here as "Fourier-dual" to distinguish it from the term dual (used for the space of continuous linear forms on a given space) which is often used to translate the Russian "sopryazhennyi" (= conjugate). (Translator's note).

$\mu(\xi)$ and $\omega(\eta)$ which occur in these equations are mutually inverse, i.e. $\mu[\omega(\eta)] = \eta$, $\omega[\mu(\xi)] = \xi$, then the corresponding functions $M(x)$ and $\Omega(y)$ will be said to be ***dual in the sense of Young***. In this case the geometrically obvious Young inequality

$$xy \leqslant M(x) + \Omega(y) \tag{1}$$

holds for any $x \geqslant 0$, $y \geqslant 0$ (cf. Fig. 1). For any x one can find a $y = y(x)$ which with the given x turns the inequality (1) into an equality. Obviously, one has to take $\mu(x)$ as this y.

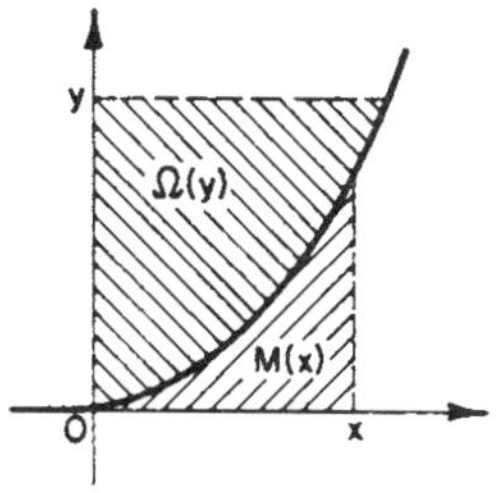

FIG. 1.

The following pairs of functions are examples of functions which are dual (in the sense of Young):

$$M(x) = x^p/p, \qquad \Omega(y) = y^q/q, \qquad \frac{1}{p} + \frac{1}{q} = 1; \tag{2}$$

$$M(x) = (x + 1)\ln(x + 1) - x, \qquad \Omega(y) = e^y - y - 1. \tag{3}$$

We leave it to the reader to verify the duality.

Let us note a few properties of dual functions which will be useful for the following.

Lemma. *If $M(x)$ is dual to $\Omega_1(y)$, $M_1(x)$ is dual to $\Omega(y)$, and $M(x) < M_1(x)$ for sufficiently large values of x, then $\Omega_1(y) > \Omega(y)$ for sufficiently large y.*

Proof. Assume that the inequality $M(x) < M_1(x)$ holds for $x > x_0$. Let us determine the $y = y(x) > y(x_0)$ for which the equality

$$xy = M_1(x) + \Omega(y)$$

holds. For the same x and y

$$xy = M(x) + \Omega_1(y).$$

Hence

$$M(x) - M_1(x) + \Omega_1(y) - \Omega(y) \geqslant 0,$$
$$0 \leqslant M_1(x) - M(x) \leqslant \Omega_1(y) - \Omega(y),$$

and consequently, for $y > y(x_0)$ the required inequality $\Omega_1(y) > \Omega(y)$ holds.

3.2. Duality Theorems for the Spaces $W_{M,a}$ and $W^{\Omega,b}$

Theorem 1. *Let $M(x)$ and $\Omega(y)$ be a pair of functions which are dual in the sense of Young. Then*

$$\widetilde{W_{M,a}} \subset W^{\Omega,1/a}. \tag{1}$$

i.e., the Fourier-dual of $W_{M,a}$ is included in the space $W^{\Omega,1/a}$.

Proof. For $|\xi| \to \infty$ the function $e^{M(a\xi)}$ increases faster than $e^{\lambda\xi}$ for any $\lambda > 0$. Therefore for $|x| \to \infty$ the test functions $\varphi(x) \in W_{M,a}$ decrease faster than $e^{-\lambda x}$ for any $\lambda > 0$. Consequently, the Fourier transform

$$\psi(\sigma) = \int_{-\infty}^{\infty} \varphi(x) e^{i\sigma x}\, dx$$

of the function $\varphi(x) \in W_{M,a}$ can be extended to complex values of $s = \sigma + i\tau$ according to the definition

$$\psi(\sigma + i\tau) = \int_{-\infty}^{\infty} \varphi(x) e^{i(\sigma+i\tau)x}\, dx, \tag{2}$$

where the integrals remain absolutely convergent.

The function $\psi(s)$ is differentiable for any s. Indeed, after a formal differentiation with respect to s, the integral in (2) becomes

$$\int_{-\infty}^{\infty} ix\varphi(x) e^{i(\sigma+i\tau)x}\, dx;$$

but according to the result of Section 2.1.2 the function $x\varphi(x)$ also belongs to the space $W_{M,a}$, so that the integral remains absolutely convergent; this guarantees the existence of the derivative of the function $\psi(s)$ on the basis of the well-known rules for differentiating improper integrals depending on a parameter.

Thus $\psi(s)$ is an entire analytic function. We have further

$$(is)^k\psi(s) = \int_{-\infty}^{\infty} \varphi^{(k)}(x)e^{ixs}\,dx. \tag{3}$$

Since $\varphi(x) \in W_{M,a}$,

$$|\, s^k\psi(s)\,| \leqslant C_{k\delta} \int_{-\infty}^{\infty} \exp[-M[(a-\delta)x] + |\,\tau\,|\,|\,x\,|]\,dx. \tag{4}$$

We further apply the Young inequality (1), Section 3.1, replacing in it x by $\gamma|\,x\,|$ and y by $(1/\gamma)|\,\tau\,|$, with $\gamma = a - 2\delta$. The exponent in (4) can be transformed as follows:

$$-M[(a-\delta)x] + |\,x\,||\,\tau\,| \leqslant -M[(a-\delta)x] + M(\gamma x) + \Omega(\tau/\gamma),$$

and we thus get the estimate

$$|\, s^k\psi(s)\,| \leqslant C_{k\delta} \exp[\Omega(\tau/\gamma)] \int_{-\infty}^{\infty} \exp[-M[(a-\delta)x] + M[(a-2\delta)x]\,dx$$
$$\leqslant C'_{k\delta} \exp[\Omega(\tau/\gamma)],$$

due to the fact that the integral

$$\int_{-\infty}^{\infty} \exp[-M[(a-\delta)x] - M[(a-2\delta)x]]\,dx \leqslant \int_{-\infty}^{\infty} \exp[-M(\delta x)]\,dx$$

has a finite value. The quantity $1/\gamma = 1/(a-2\delta)$ can be represented in the form $(1/a) + \rho$, where ρ is arbitrarily small, together with δ. We thus obtain the result that the function $\psi(s)$ belongs to the space $W^{\Omega,1/a}$, where $\Omega(y)$ is the dual in the Young sense to the function $M(x)$. We note at the same time that the Fourier operator maps a bounded set $A \subset W_{M,a}$ into a bounded set $\tilde{A} \subset W^{\Omega,1/a}$ and hence is a *continuous* operator.

Theorem 2. *Let $M(x)$ be the function which is dual in the Young sense to $\Omega(y)$. Then*

$$\widetilde{W^{\Omega,b}} \subset W_{M,1/b}\,. \tag{5}$$

Proof. The function $\varphi(x + iy) \in W^{\Omega,b}$, vanishes faster than any power of $1/|\,x\,|$, as $|\,x\,| \to \infty$, uniformly in any strip $|\,y\,| \leqslant y_0$, and in this strip the absolute value of the function $e^{i\sigma z}$ remains bounded. Therefore using Cauchy's theorem in the expression for the Fourier transform

$$\psi(\sigma) = \int_{-\infty}^{\infty} \varphi(x)e^{i\sigma x}\,dx,$$

one may replace the integration path by any horizontal straight line without modifying the result

$$\psi(\sigma) = \int_{-\infty}^{\infty} \varphi(x + iy)e^{i(x+iy)\sigma}\, dx. \tag{6}$$

Differentiating (6) with respect to σ we obtain

$$\psi^{(q)}(\sigma) = \int_{-\infty}^{\infty} (iz)^q \varphi(z) e^{i\sigma z}\, dx, \tag{7}$$

where the differentiation under the integral sign is legitimate due to the absolute convergence of the resulting integral. It follows from (7) that

$$|\, \psi^{(q)}(\sigma) \,| \leqslant \int_{-\infty}^{\infty} |\, z^q \varphi(z) \,|\, e^{-\sigma y}\, dx \leqslant e^{-\sigma y} \int_{-\infty}^{\infty} \frac{|\, z\,|^{q+2} + |\, z\,|^q}{x^2+1} \,|\, \varphi(z) \,|\, dx,$$

by making use of the obvious inequality

$$|\, z\,|^q \leqslant \frac{|\, z\,|^{q+2} + |\, z\,|^q}{x^2+1};$$

further, making use of the definition of $\varphi(z)$, we obtain the estimate

$$|\, \psi^{(q)}(\sigma) \,| \leqslant e^{-\sigma y}[C_{p,q+2} + C_{pq}]\, e^{\Omega[(b+\rho)y]} \leqslant C'_{pq}\, e^{-\sigma y}\, e^{\Omega[(b+\rho)y]}. \tag{8}$$

Until now y has been an arbitrary number. Let us now choose the sign of y in such a manner that the equality $\sigma y = |\, \sigma \,|\, |\, y\,|$ be satisfied, and the absolute value of y so that the Young inequality (1) in Section 3.1, with y replaced by $(b+\rho)|\, y\,|$ and x replaced by $|\, \sigma\,|/(b+\rho)$, becomes the equality

$$|\, \sigma \,|\, |\, y\,| = M\left(\frac{|\, \sigma\,|}{b+\rho}\right) + \Omega[(b+\rho)\,|\, y\,|].$$

Then the exponent in (8) becomes

$$-\sigma y + \Omega[(b+\rho)y] = -\,|\, \sigma \,|\, |\, y\,| + \Omega[(b+\rho)\,|\, y\,|] = -M\left(\frac{|\, \sigma\,|}{b+\rho}\right).$$

Replacing $1/(b+\rho)$ by $(1/b) - \delta$, where δ is arbitrarily small together with ρ, we obtain the inequality

$$|\, \psi^{(q)}(\sigma) \,| \leqslant C_{\delta q}\, e^{-M\left[\left(\frac{1}{b} - \delta\right) x\right]},$$

i.e., $\psi(\sigma) \in W_{M,1/b}$.

Combining Theorems 1 and 2 and taking into account the fact that $\tilde{\tilde{\varphi}}(x) = \varphi(-x)$ we obtain the following fundamental result:

Theorem 3. *If the functions $M(x)$ and $\Omega(y)$ are mutually dual in the sense of Young, then*

$$\widetilde{W_{M,a}} = W^{\Omega,1/a}, \qquad \widetilde{W^{\Omega,b}} = W_{M,1/b}. \tag{9}$$

The Fourier operator which maps $W^{\Omega,b}$ onto $W_{M,1/b}$ is also a *continuous* operator. This can be proved directly, or by making use of the one-to-one character of the mapping together with the theorem on the continuity of the inverse operator (Volume 2, Chapter I, Section 7).

In addition to Eqs. (9) *the following Fourier-duality relations hold for the spaces M_M and W^Ω:*

$$\widetilde{W_M} = W^\Omega, \qquad \widetilde{W^\Omega} = W_M, \tag{10}$$

and the Fourier operator is continuous in this case too.

If convenient, the sign $\sim$ in Eqs. (9) and (10) can be understood as the inverse Fourier transform, rather than the direct one, since each of the W spaces contains the function $\varphi(-x)$ together with the function $\varphi(x)$. The same is also true for the relations which will be derived in the following section.

3.3. Duality Theorems for the Spaces $W_{M,a}^{\Omega,b}$

Theorem 4. *Let $\Omega_1(y)$ and $M_1(x)$ be the functions which are dual in the sense of Young to the functions $M(x)$ and $\Omega(y)$, respectively. Then*

$$\widetilde{W_{M,a}^{\Omega,b}} = W_{M_1,1/b}^{\Omega_1,1/a}. \tag{1}$$

Proof. In the expression for the Fourier transform of the function $\varphi(x) \in W_{M,a}^{\Omega,b}$

$$\psi(\sigma) = \int_{-\infty}^{\infty} \varphi(x) e^{i\sigma x}\, dx$$

one can, due to the analyticity of $\varphi(x + iy)$ and its exponential decrease along the real axis in the z-plane (which is uniform in any horizontal strip), shift the integration contour into any horizontal straight line, as was done in Theorem 2; on the other hand, one can carry out an analytic continuation, replacing σ in both sides of the equality by $s = \sigma + i\tau$, as was done in Theorem 1. This results in

$$\psi(\sigma + i\tau) = \int_{-\infty}^{\infty} \varphi(x + iy) \exp[i(x + iy)(\sigma + i\tau)]\, dx.$$

An estimate of the absolute value yields

$$\begin{aligned} |\psi(\sigma + i\tau)| &\leqslant C_{\delta\rho} \int_{-\infty}^{\infty} \exp[-M[(a-\delta)x] + \Omega[(b + \rho)y]] \exp[-y\sigma - x\tau]\, dx \\ &\leqslant C_{\delta\rho} \exp[-\sigma y + \Omega[(b + \rho)y]] \\ &\quad \times \int_{-\infty}^{\infty} \exp[-M[(a - \delta)x] + |x||\tau|]\, dx. \end{aligned}$$

Carrying out the same transformations of the integral as were used in Theorem 1, the integral will be replaced by the larger quantity

$$C_\sigma' e^{\Omega_1(\tau/a-2\delta)}.$$

Choosing y as in Theorem 2, and carrying out the same transformations of the factor in front of the integral, as in Theorem 2, this factor will be replaced by $\exp[-M_1\sigma/(b + \rho)]$. Setting

$$\frac{1}{a - 2\delta} = \frac{1}{a} + \rho_1, \qquad \frac{1}{b + \rho} = \frac{1}{b} - \delta_1,$$

where δ_1 and ρ_1 are arbitrarily small together with δ and ρ, we obtain the estimate

$$|\psi(\sigma + i\tau)| \leqslant C'_{\rho_1\delta_1} \exp\left[-M_1\left[\left(\frac{1}{b} - \delta_1\right)\sigma\right] + \Omega_1\left[\left(\frac{1}{a} + \rho_1\right)\tau\right]\right],$$

where $C'_{\rho_1\delta_1} = C_{\rho\delta}C_\delta'$. Thus $\psi(\sigma) \in W_{M_1,1/b}^{\Omega_1,1/a}$; at the same time it is obvious that the Fourier transform $\psi = \tilde{\varphi}$ is a bounded operator which maps $W_{M,a}^{\Omega,b}$ into $W_{M_1,1/b}^{\Omega_1,1/a}$.

Applying the same reasoning to the space $W_{M_1,1/b}^{\Omega_1,1/a}$ we find that it is mapped by the Fourier transformation into the space $W_{M,a}^{\Omega,b}$. Due to the uniqueness of the Fourier transform both mappings are one-to-one. One thus obtains the assertion of the theorem.

Together with Eq. (1), the duality relation for the space W_M^Ω is also true:

$$\widetilde{W_M^\Omega} = W_{M_1}^{\Omega_1}, \tag{2}$$

where the Fourier transform is also a continuous operator.

Examples. Defining

$$M(x) = \frac{x^p}{p}, \qquad \Omega_1(y) = \frac{y^q}{q}, \qquad M_1(x) = \frac{x^r}{r},$$

$$\Omega(y) = \frac{y^s}{s} \qquad \left(\frac{1}{p} + \frac{1}{q} = \frac{1}{r} + \frac{1}{s} = 1\right)$$

and, for simplicity of notation, replacing the symbols of the functions M, Ω_1, M_1, Ω by the respective powers p, q, r, s (as indicated at the end of Section 1.3), we obtain the relations

$$\widetilde{W_{p,a}} = W^{q,1/a}, \qquad \widetilde{W_p} = W^q, \tag{3}$$

$$\widetilde{W^{r,b}} = W_{s,1/b}, \qquad \widetilde{W^r} = W_s, \tag{4}$$

$$\widetilde{W^{r,b}_{p,a}} = W^{q,1/a}_{s,1/b}, \qquad \widetilde{W^r_p} = W^q_s. \tag{5}$$

4. The Case of Several Variables

4.1. Definitions of Test Function Spaces

Let the functions $M_1(x_1),\ldots, M_n(x_n)$ (each depending on one variable) be given; assume they satisfy the conditions given in Section 1. By definition the space $W_{M_1,\ldots,M_n} = W_M$ consists of all infinitely differentiable functions $\varphi(x) = \varphi(x_1,\ldots, x_n)$ which satisfy the inequalities

$$|D^q\varphi(x)| \equiv \left|\frac{\partial^{q_1+\cdots+q_n}\varphi(x)}{\partial x_1^{q_1}\cdots\partial x_n^{q_n}}\right| \leqslant C_q \exp[-M_1(a_1x_1) - \cdots - M_n(a_nx_n)]. \tag{1}$$

A sequence $\varphi_\nu(x) \in W_M$ is said to converge to zero if: (a) the functions $\varphi_\nu(x)$ and their derivatives of any order converge to zero uniformly in any bounded region (regular convergence) and, (b) in the inequalities (1) for the functions one may choose the constants C_q, $a_1,\ldots, a_n$ independently of ν (uniform boundedness).

Let $\Omega_1(y_1),\ldots, \Omega_n(y_n)$ be analogous functions of $y_1,\ldots, y_n$; by definition, the space $W^{\Omega_1,\ldots,\Omega_1} = W^\Omega$ consists of all entire analytic functions of the n complex variables $z_j = x_j + iy_j$, for which

$$|z^k\varphi(z)| \equiv |z_1^{k_1}\cdots z_n^{k_n}\varphi(z_1,\ldots, z_n)| \leqslant C_k \exp[\Omega_1(b_1y_1) + \cdots + \Omega_n(b_ny_n)]. \tag{2}$$

A sequence $\varphi_\nu(z) \in W^\Omega$ is said to converge to zero if: (a) the functions $\varphi_\nu(z)$ converge to zero uniformly in any bounded domain of the n-dimensional complex space (regular convergence), and (b) the constants C_k, $b_1, \ldots, b_k$ in the inequalities (2) written for the function $\varphi_\nu(z)$ can be chosen independently of ν (uniform boundedness).

If both $M_j(x_j)$ and $\Omega_j(y_j)$ are given, we can also construct the space $W_M{}^\Omega = W_{M_1,\ldots,M_n}^{\Omega_1,\ldots,\Omega_n}$: it consists of entire analytic functions $\varphi(z) \equiv \varphi(z_1, \ldots, z_n)$ for which

$$|\varphi(z)| \leqslant C \exp[-M_1(a_1x_1) - \cdots - M_n(a_nx_n) + \Omega_1(b_1y_1) + \cdots + \Omega_n(b_ny_n)] \tag{3}$$

A sequence of functions $\varphi_\nu(z)$ is said to converge to zero in this space if: (a) the functions $\varphi_\nu(z)$ converge to zero uniformly in any bounded domain of the n-dimensional complex space (regular convergence), and (b) the constants $C, a_1, \ldots, a_n, b_1, \ldots, b_n$ in the inequalities (3), written for the functions $\varphi_\nu(z)$, can be chosen independently of ν.

Replacing the constants a_j by $a_j - \delta_j$ and b_j by $b_j + \rho_j$ in the above inequalities keeping the constants a_j and b_j fixed, and requiring the validity of the inequalities (1)–(3) for all positive δ_j and ρ_j, we obtain the definitions of the spaces $W_{M,a}$, $W^{\Omega,b}$, and $W_{M,a}^{\Omega,b}$; these are (countably normed) perfect topological vector spaces and their unions yield the appropriate spaces W_M, W^Ω, and $W_M{}^\Omega$.

4.2. Operations in Test Function Spaces

In all the indicated spaces the operations of differentiation $\partial/\partial x_j$ and multiplication by x_j (or $\partial/\partial z_j$ and z_j, respectively) are defined and continuous. The proof is the same as in Section 2; in making use of the Cauchy formula the integration contour is chosen in the appropriate $x_j + iy_j$ plane.

If the entire analytic function $f(z)$ satisfies the inequality

$$|f(z_1, \ldots, z_n)| \leqslant C \exp[\Omega_1(b_1{}^0y_1) + \cdots + \Omega_n(b_n{}^0y_n)](1 + |x_1|^{h_1}) \cdots (1 + |x_n|^{h_n}),$$

then it is a multiplier in the space W^Ω, and for any function $\varphi \in W^{\Omega,b}$ the product $f\varphi$ belongs to the space $W^{\Omega,b+b^0}$

If the analytic function $f(z)$ satisfies the inequality

$$|f(z_1, \ldots, z_n)| \leqslant C \exp[-M_1(a_1{}^0x_1) - \cdots - M_n(a_n{}^0x_n) + \Omega_1(b_1{}^0y_1) + \cdots + \Omega_n(b_n{}^0y_n)],$$

then for any function $\varphi \in W_{M,a}^{\Omega,b}$ with $a_j > a_j^0$ $(j = 1, 2, \ldots, n)$ the product $f\varphi$ belongs to the space $W_{M,a-a^0}^{\Omega,b+b^0}$.

In particular, the function $f(z) = \exp[i(\sigma, z)] = \exp[i(\sigma_1 z_1 + \cdots + \sigma_n z_n)]$ is a multiplier in any of the spaces $W_{M,a}^{\Omega,b}$ for any real $\sigma_1, \ldots, \sigma_n$. All these assertions can be proved exactly in the same manner as the analogous propositions were proved in Section 2 for functions of one variable.

4.3. Duality Theorems

The Fourier transform

$$\psi(\sigma_1, \ldots, \sigma_n) = \int_{-\infty}^{\infty} \cdots \int \varphi(x_1, \ldots, x_n) \exp[i(x_1\sigma_1 + \cdots + x_n\sigma_n)]\, dx_1 \cdots dx_n \tag{1}$$

of the function $\varphi(x)$ is a bounded operator which takes the space

$$W_{M,a} \text{ into } W^{\Omega_1, 1/a} \qquad \left(\frac{1}{a} = \left(\frac{1}{a_1}, \ldots, \frac{1}{a_n}\right)\right),$$

the space

$$W^{\Omega,b} \text{ into } W_{M_1,1/b} \qquad \left(\frac{1}{b} = \left(\frac{1}{b_1}, \ldots, \frac{1}{b_n}\right)\right),$$

and the space

$$W_{M,a}^{\Omega,b} \text{ into } W_{M_1,1/b}^{\Omega_1,1/a}.$$

Here M_1 and Ω_1 are functions which are dual in the sense of Young, to the functions Ω and M, respectively. This assertion can be proved by using the method of Section 3 and making use of the fact that after applying the estimate (1) (or (2) or (3), respectively) to the function $\varphi(x)$, the integrand in (4) becomes a product of n factors, each depending on only one coordinate z_j and one coordinate s_j. Therefore the integral becomes a product of n factors, each of which can be treated according to the methods of Section 3.

4.4. On the Nontriviality and "Richness" in Functions of the Test Function Spaces

The problem of nontriviality of the spaces W_M^{Ω} can be reduced to the corresponding problem for each of the spaces $W_{M_j}^{\Omega_j}$ with j fixed. If all these spaces are nontrivial, and if $\varphi_j \in W_{M_j}^{\Omega_j}$ is a nonvanishing function, then the product $\prod_{j=1}^{n} \varphi_j$ will also be nonvanishing and will belong to the space $W_M{}^\Omega$. If even for one j the space $W_{M_j}^{\Omega_j}$ is trivial, the whole space W_M is trivial, since any function $\varphi(z) \in W_M{}^\Omega$ considered for fixed z_k $(k \neq j)$ as a function of z_j is a member of the space $W_{M_j}^{\Omega_j}$. In particular, the space $W_M{}^\Omega$ will be nontrivial if the sufficient condition of nontriviality, formulated at the end of Section 1.4, is satisfied for any pair M_j, Ω_j. All the results of Section 1 referring to the denseness of the space $W_M{}^\Omega$ in the normed space E can be formulated in the n-dimensional case in an obvious fashion.

CHAPTER II

UNIQUENESS CLASSES FOR THE CAUCHY PROBLEM

1. Introduction

In this, and the following chapters, we treat some applications of the theory of generalized functions to general existence and uniqueness problems for linear partial differential equations.

We restrict our attention to the investigation of the Cauchy problem, mainly for equations with coefficients which do not depend on the spatial variables. We shall not be concerned with other types of problems: boundary value problems, mixed problems with general boundary conditions, etc., since the methods of generalized functions are not yet sufficiently developed in this direction.[1] We hope however, that the methods which have been developed for the investigation and solution of the Cauchy problem will turn out to be useful in the future also for the solution of other kinds of problems.

We shall consider systems of partial differential equations of the form

$$\frac{\partial u_j(x, t)}{\partial t} = \sum_{k=1}^{m} P_{jk}\left(\frac{\partial}{\partial x}\right) u_k(x, t) \qquad (j = 1, \ldots, m), \tag{1}$$

where $P_{jk}(s)$ are polynomials in s with constant coefficients.[2] To this form one can reduce the formally more complicated systems

$$\frac{\partial^{\mu_j} u_j(x, t)}{\partial t^{\mu_j}} = \sum_{k=1}^{m} P_{jk}\left(\frac{\partial}{\partial t}, \frac{\partial}{\partial x_1}, \ldots, \frac{\partial}{\partial x_n}\right) u_k(x, t), \tag{2}$$

where the order in $\partial/\partial t$ of the polynomial P_{jk} is smaller than μ_j. The wave equation, for instance, belongs to this type.

[1] *Translator's note:* Cf., however, the book by Hörmander, quoted in the translator's preface.

[2] In the appendix to this chapter various special cases are considered which do not fit directly into the general framework (1), namely systems with convolution operators, in particular difference-differential equations and equations with variable coefficients.

In order to obtain the system (1) from the system (2) one introduces new unknown functions according to the relations

$$\begin{array}{lll} u_{11} = u_1, & u_{12} = \dfrac{\partial u_1}{\partial t}, \ldots, u_{1\mu_1} & = \dfrac{\partial^{\mu_1 - 1} u_1}{\partial t^{\mu_1 - 1}}, \\ \cdots & \cdots & \cdots \\ \cdots & \cdots & \cdots \\ u_{m1} = u_m, & u_{m2} = \dfrac{\partial u_m}{\partial t}, \ldots, u_{m,\mu_m - 1} & = \dfrac{\partial^{\mu_m - 1} u_m}{\partial t^{\mu_m - 1}}. \end{array}$$

The Cauchy problem for the system (1) consists in determining a solution which satisfies the initial conditions

$$u_j(x, 0) = u_j(x) \qquad (j = 1, \ldots, n). \tag{3}$$

where $u_j(x)$ are given functions.

We will be interested in the construction of *uniqueness classes* for solutions of the Cauchy problems, as well as in the construction of *correctness classes.*

A *uniqueness class* is a (linear topological) space of (ordinary or generalized) functions of the argument x, for which the uniqueness of the solution of the Cauchy problem is guaranteed for given initial conditions, provided the solution exists.

A *correctness class* is the totality of *ordinary* functions of the arguments x, for which the existence of the solution of the Cauchy problem is guaranteed for an arbitrary choice of initial conditions (again within a given class of functions), as well as its uniqueness and continuous dependence of the solution on the initial conditions.

In this and the following chapter we shall make use of generalized functions defined on test function spaces of infinitely differentiable or analytic functions with definite conditions on their behavior at infinity. In particular, we shall make use of spaces of type S and W. It is obvious that for the treatment of boundary value problems one will have to introduce test function spaces which are defined so as to take into account the boundary conditions.

The method of Fourier transforms plays an essential role in this theory. Until recently it was believed that this method is applicable only to problems involving integrable functions.[3] The theory of generalized functions makes it possible to construct Fourier transforms for any functions, with arbitrary growth properties, and this allows one to make full use of the Fourier transform.

[3] Or with functions which become integrable after multiplication with e^{-px} (a Laplace transform which is a modification of the Fourier transform).

A second essential device is the use of differential operators of infinite order. Consider, for example, the solution of the Cauchy problem for the equation

$$\frac{\partial u(x, t)}{\partial t} = P\left(\frac{\partial}{\partial x}\right) u(x, t),$$

with the initial condition

$$u(x, 0) = u_0(x),$$

which can be formally written in the form

$$u(x, t) = \exp[tP(\partial/\partial x)]\, u_0(x),$$

i.e., as the result of applying the differential operator of infinite order

$$\exp[tP(\partial/\partial x)] = \sum_{n=0}^{\infty} \frac{t^n P^n(\partial/\partial x)}{n!}$$

to the initial function $u_0(x)$. Obviously it is necessary to give a precise definition of this operator and to specify exactly under what conditions it is applicable.

Consider as an example the Cauchy problem for the equation

$$\frac{\partial u(x, t)}{\partial t} = a \frac{\partial^2 u(x, t)}{\partial x^2}, \tag{4}$$

where a is a complex constant. It turns out that for any a the uniqueness class of solutions of the Cauchy problem is always the same, namely the class of functions which for fixed t satisfy the inequality

$$|f(x)| \leqslant C_1 \exp[C_2 x^2]. \tag{5}$$

However, whereas for the heat equation ($a = 1$) the same set of functions also serves as the correctness class of the problem, for the Schrödinger equation ($a = i$) the correctness class will consist of the totality of sufficiently smooth functions (which in addition increase at most like a power of x) and for the inverse heat equation ($a = -1$) the correctness class will consist only of the totality of (entire) analytic functions of a given order (of growth).

This example shows that the problem of finding the uniqueness classes of solutions of the Cauchy problem is distinct from the problem of finding the correctness classes and has to be treated separately, using an independent method. This is the reason for separating the problem of uniqueness classes into an independent chapter.

We will show in this chapter that for each system (1) (consequently, also for systems of the form (2)) there exists a uniqueness class of solutions of the Cauchy problem, described by inequalities of the form

$$|f(x)| \leqslant C \exp[b\,|\,x\,|^r],$$

where the exponent r ($\leqslant \infty$) depends on the specific system.

We shall make use of the well-known device, which it seems, is due to Holmgren (1901), consisting of deriving the uniqueness of the solution of the adjoint equation from the existence of the solutions of a given equation for arbitrary initial conditions.

In the following section this trick is investigated in its abstract form. The solution of a given equation will be considered as a generalized function on a certain test function space. Depending on the equation, this test function space has then to be restricted so as to guarantee that the Cauchy problem for the adjoint equation is always solvable in this space.

2. The Cauchy Problem in a Topological Vector Space

2.1. The Connection between the Solutions of the Cauchy Problem in a Given Space and in the Dual Space

Let A_t be a given linear continuous operator which depends continuously on the parameter t, and which for each t ($0 \leqslant t \leqslant T$) maps the topological vector space Φ (on which A_t is defined) into itself. The adjoint operator A_t^* is defined on the dual space Φ' and maps this space into itself. Consider the differential equation

$$du(t)/dt = -A_t^* u(t), \tag{1}$$

where $u(t)$ is an unknown element of the space Φ'. The problem of finding a solution of this equation, satisfying the initial condition

$$u(0) = u_0 \in \Phi' \tag{2}$$

will be called the abstract Cauchy problem. The solution of the Cauchy problem (1)–(2) is closely connected with the solution of an analogous Cauchy problem in the space Φ, namely with the problem of finding the solution of the equation

$$d\varphi(t)/dt = A_t \varphi(t), \tag{3}$$

with $\varphi(t) \in \Phi$, and satisfying the initial condition

$$\varphi(t_0) = \varphi_0 \in \Phi. \tag{4}$$

The connection between the two problems consists in the following.

For brevity we shall say that the Cauchy problem (3)–(4) is *always solvable* if for each t_0, $0 \leqslant t_0 \leqslant T$, and each element φ_0 there exists a solution $\varphi(t)$ defined for all $0 \leqslant t \leqslant T$, which for $t = t_0$ becomes equal to φ_0, and is such that $\varphi(t)$ depends linearly and continuously on φ_0. The following theorem holds.

Theorem. *If the Cauchy problem* (3)–(4) *is always solvable in the* (*test function*) *space* Φ, *then the Cauchy problem* (1)–(2) *in the dual space* Φ' *has a solution for any initial functional* u_0; *this solution is unique in* Φ' *and depends continuously on the functional* u_0 *in the sense of the topology of the space* Φ'.[4]

Before proving this theorem, let us carry out some preliminary constructions.

We denote by $Q_{t_0}^t$ the linear continuous operator in Φ which maps the vector φ_0 into the solution $\varphi(t)$ of the Cauchy problem (3)–(4). By assumption, this operator is defined for any values of t_0 and t within the interval $[0, T]$, and has the properties

$$\begin{aligned} \frac{d[Q_{t_0}^t \varphi(t_0)]}{dt} &= A_t Q_{t_0}^t \varphi(t_0), \\ Q_{t_0}^{t_0} \varphi(t_0) &= \varphi(t_0). \end{aligned} \tag{5}$$

The last equation implies

$$Q_{t_0}^t = E \text{ (the unit operator) for any } t_0 . \tag{6}$$

It follows from Eq. (5) that the operator $Q_{t_0}^t$ is a solution of the equation

$$\frac{dQ_{t_0}^t}{dt} = A_t Q_{t_0}^t . \tag{7}$$

In particular, we obtain

$$\left. \frac{dQ_{t_0}^t}{dt} \right|_{t=t_0} = A_{t_0} . \tag{8}$$

[4] Φ' is endowed with the weak topology; consequently $du(t)/dt$ is a weak derivative (cf. the beginning of Chapter I, Volume 2).

For three arbitrary values t_0, t_1, t_2 in the interval $[0, T]$ one can write

$$\varphi(t_2) = Q_{t_1}^{t_2}\varphi(t_1) = Q_{t_1}^{t_2}Q_{t_1}^{t_1}\varphi(t_0),$$

but, on the other hand,

$$\varphi(t_2) = Q_{t_0}^{t_2}\varphi(t_0).$$

Comparing these two equations we obtain

$$Q_{t_1}^{t_2}Q_{t_0}^{t_1} = Q_{t_0}^{t_2}. \tag{9}$$

Setting $t_2 = t_0$ in this equation we obtain

$$Q_{t_1}^{t_0}Q_{t_0}^{t_1} = Q_{t_0}^{t_0} = E. \tag{10}$$

Replacing t_1 by t and differentiating with respect to it, we have

$$\frac{dQ_t^{t_0}}{dt}Q_{t_0}^t + Q_t^{t_0}\frac{dQ_{t_0}^t}{dt} = 0.$$

Making use of Eq. (7) we find:

$$\frac{dQ_t^{t_0}}{dt}Q_{t_0}^t + Q_t^{t_0}A_tQ_{t_0}^t = 0,$$

whence, by right multiplication with $Q_{t_0}^t$, we finally arrive at

$$\frac{dQ_t^{t_0}}{dt} = -Q_t^{t_0}A_t. \tag{11}$$

For the adjoint operators we thus obtain

$$\frac{dQ_t^{t_0 *}}{dt} = -A_t^* Q_t^{t_0 *}. \tag{12}$$

We are now ready to prove the theorem. Let $t_0 = 0$, $u(t) = Q_t^{0*}u_0$; then the functional $u(t)$ is a solution of the Cauchy problems (1) and (2). Indeed, applying the operator equation (12) to the vector u_0 we have

$$\frac{du(t)}{dt} = -A_t^* u(t),$$

so that Eq. (1) is satisfied. Since for $t = 0$ the operator $Q_t{}^{0*}$ becomes $E^* = E$, the initial condition is also satisfied. We note finally that the functional $u(t)$ depends continuously on the initial element u_0, due to the continuity of the operator Q_t^{0*}. Let us prove now that the Cauchy problem can have only a unique solution in the space Φ'. For this it is sufficient to show that the only solution of the equation (1) with the initial condition $u_0 = 0$ is the functional $u(t) \equiv 0$. For the proof, we select an arbitrary t_0, $0 \leqslant t_0 \leqslant T$, and apply the functional $u(t)$ to the test function $Q_{t_0}^t \varphi_0$, where φ_0 is an arbitrary element of the space Φ. Differentiating the result with respect to t and making use of Eqs. (1) and (6), we obtain

$$\begin{aligned}\frac{d}{dt}(u(t), Q_{t_0}^t \varphi_0) &= \left(\frac{du(t)}{dt}, Q_{t_0}^t \varphi_0\right) + \left(u(t), \frac{dQ_{t_0}^t}{dt}\varphi_0\right)\\ &= (-A_t{}^*u(t), Q_{t_0}^t \varphi_0) + (u(t), A_t Q_{t_0}^t \varphi_0)\\ &= (-A_t{}^*u(t), Q_{t_0}^t \varphi_0) + (A_t{}^*u(t), Q_{t_0}^t \varphi_0) = 0.\end{aligned}$$

It thus follows that $(u(t), Q_{t_0}^t \varphi_0)$ is a constant. The initial condition $u(0) = 0$ implies that this quantity vanishes for all t. In particular, for $t = t_0$, it follows

$$(u(t_0), \varphi_0) = 0.$$

Since φ_0 is an arbitrary element of Φ, the functional $u(t_0)$ is the null-functional. From the arbitrary choice of t_0 between 0 and T, it then follows that $u(t) \equiv 0$, as expected.

2.2. A More General Uniqueness Theorem

If one is interested only in the problem of uniqueness, the situation described in Section 2.1 can be considerably generalized. We assume now that the space Φ is part of a larger space Φ_1, which in turn is contained in a third space E:

$$\Phi \subset \Phi_1 \subset E.$$

Each of these inclusions by assumption preserves linear operations as well as convergence, i.e., $\varphi_n \to 0$ in Φ implies $\varphi_n \to 0$ in Φ_1 and $\varphi_n \to 0$ in Φ_1 implies $\varphi_n \to 0$ in E. We further assume that the space Φ is dense in E. Every linear continuous functional defined on E is thus at the same time a continuous linear functional on Φ and Φ_1. We further assume that the operator A_t can be extended from the space Φ, on which it is originally defined, to the space Φ_1.

Theorem. *If for any t_0 and t $(0 \leqslant t \leqslant t_0 \leqslant T)$ a linear continuous operator $Q_{t_0}^t$ is defined on the space Φ which maps this space into the space Φ_1 and in addition, if applied to an arbitrary vector $\varphi_0 \in \Phi$ yields a solution of the Cauchy problem*

$$\frac{d\varphi(t)}{dt} = A_t\varphi(t), \qquad \varphi(t_0) = \varphi_0\,, \qquad \varphi(t) \in \Phi_1\,, \tag{1}$$

then the Cauchy problem

$$\frac{du(t)}{dt} = -A_t{}^*u(t), \qquad u(0) = u_0 \tag{2}$$

admits in the interval $0 \leqslant t \leqslant T$ the unique solution[5] *$u(t) \in E'$.*

Proof. One can reason in the same manner as for the Theorem in Section 2.1. The operator $Q_{t_0}^t$ satisfies the equation

$$\frac{dQ_{t_0}^t}{dt} = A_t Q_{t_0}^t.$$

Only this equation is used in the proof of the uniqueness of the solution of the Cauchy problem (1) and (2) at the end of the preceding section, so that we can use the essential idea of that proof. We apply the functional $u(t) \in E'$, which is a solution of the Cauchy problem (1) and (2) with initial condition $u(0) = 0$, to the vector $Q_{t_0}^t\varphi_0 \in \Phi_1 \subset E$. Thus, in the same manner as in Section 2.1, we show that $(u(t_0), \varphi_0) = 0$, i.e., the functional $u(t_0)$ vanishes on the space Φ. Since by assumption $u(t_0)$ is defined on all of E, in which Φ is assumed to be dense, it follows that $u(t_0)$ is the null-functional of E.

3. The Cauchy Problem for Systems of Partial Differential Equations. The Operator Method

3.1. Introduction

We consider the system of partial differential equations

$$\frac{\partial u_j(x, t)}{\partial t} = \sum_{k=1}^{m} P_{jk}\left(i\frac{\partial}{\partial x}\right) u_k(x, t) \qquad (j = 1, 2, \ldots, m), \tag{1}$$

[5] We note that in order to conclude that the solution of the Cauchy problem (2) is unique in the interval $0 \leqslant t \leqslant T$ for a given initial condition at $t = 0$, it is sufficient to assume that the operator $Q_{t_0}^t$ exists only for $0 \leqslant t < t_0 \leqslant T$ (i.e., the upper index is not larger than the lower one).

Here, the unknown functions are $u_j(x, t)$. The differential operators $P_{jk}(i(\partial/\partial x))$ are polynomials in the derivatives $\partial/\partial x_1, \ldots, \partial/\partial x_n$ of maximal order p (in the ensemble of arguments); the number p will be called the *order of the system* (1).

The *Cauchy problem* consists in determining the set of functions $u_j(x, t)(j = 1, 2, \ldots, m)$ which for $t \geqslant 0$ satisfy the Eqs. (1) and for $t = 0$ satisfy the initial data

$$u_j(x, 0) = u_j(x), \tag{2}$$

where $u_j(x)$ are given functions. The ensemble $\{u_j(x, t)\}$ is called *a solution* of the problem (1) and (2).

It is often more convenient to rewrite the Cauchy problem (1) and (2) in vector form

$$\frac{\partial u(x, t)}{\partial t} = P\left(i \frac{\partial}{\partial x}\right) u(x, t), \tag{3}$$

$$u(x, 0) = u(x), \tag{4}$$

with $u(x, t)$ an unknown vector-function, $u(x, t) = \{u_j(x, t)\}$ and $P(i(\partial/\partial x))$ is a matrix (m rows, m columns) consisting of differential operators.

We are interested in finding the uniqueness classes of solutions of the Cauchy problem (1) and (2), i.e., to indicate the class of functions $u(x)$ such that any solution, if it exists, and belongs to this class for any t $(0 \leqslant t \leqslant T)$, is certainly unique.

Before attacking this problem in general, we illustrate the basic idea on the example of the equation

$$\frac{\partial u(x, t)}{\partial t} = a \frac{\partial^2 u(x, t)}{\partial t^2}, \tag{5}$$

where a is an arbitrary (complex) constant.

For each value of t the unknown $u(x, t)$ is a generalized function over a test function space Φ. In agreement with the plan outlined in Section 1, this test function space should be so chosen as to guarantee the possibility of solving the Cauchy problem for the corresponding (inverse-conjugate) equation

$$\frac{\partial \varphi(x, t)}{\partial t} = -\bar{a} \frac{\partial^2 \varphi(x, t)}{\partial x^2}$$

Formally, the solution can be written in the form

$$\varphi(x, t) = \exp\left[-\bar{a}(t - t_0) \frac{d^2}{dx^2}\right] \varphi_0(x),$$

and the space Φ should be so chosen that the differential operator of infinite order

$$\underset{\sim}{Q}_{t_0}^{t} = \exp\left[-\bar{a}(t - t_0)\frac{d^2}{dx^2}\right]$$

be meaningful in Φ.

Furthermore, it is convenient to select the space Φ as small as practicable in order that the uniqueness theorem be valid in as wide a class of generalized functions as possible. Let us choose the space Φ among the spaces S_α^β, described in Chapter IV, Volume 2. The space S_α^β consists of functions $\varphi(x)$ satisfying the inequalities

$$| x^k\varphi^{(q)}(x) | \leqslant CA^kB^qk^{k\alpha}q^{q\beta}.$$

The function $f(s) = \exp(-\bar{a}(t - t_0)s^2)$ with

$$f\left(\frac{d}{dx}\right) = \exp\left[-\bar{a}(t - t_0)\frac{d^2}{dx^2}\right]$$

is of order 2. In order to be able to apply the fundamental theorem of Section 5, Chapter IV, Volume 2, one must consider a space S_α^β with $1/\beta > 2$ or $\beta > \frac{1}{2}$. We further choose α as small as possible, but such that the space S_α^β is nontrivial. This α will be $1 - \beta$, i.e., larger than $\frac{1}{2}$. The test functions of this space exhibit for $| x | \to \infty$ an exponential decrease of order $1/\alpha$, i.e., smaller than 2. The solution of Cauchy's problem for Eq. (1) is unique in the dual space, on the basis of the results of Section 1. In particular, the problem is unique for ordinary functions $f(x)$ which grow slower than $\exp(x^2)$ for $| x | \to \infty$:

$$|f(x)| \leqslant C\exp[x^2 - \epsilon].$$

Thus, we have indicated a uniqueness class of solutions of the Cauchy problem for Eq. (5).[6] Note that this class does not depend on the value of the constant a. Even in the incorrect case, Re $a < 0$, the uniqueness of the Cauchy problem is maintained in this class of functions.

3.2. Preliminary Constructions and Formulation of the Fundamental Theorem

We now start with the examination of the general case.

We apply the method outlined in Section 2 in the following from. The elements of the vector spaces Φ, Φ_1 which occurred in the con-

[6] This class is not completely specified here. As will be seen in the sequel, the exponent $2 - \epsilon$ could be replaced by 2. This will necessitate the use of the spaces $S_{\alpha A}^{\beta B}$ instead of S_α^β.

structions of Section 2, will be m-dimensional vector functions $\varphi = (\varphi_1, \dots, \varphi_m)$, such that the components $\varphi_1(x), \dots, \varphi_m(x)$ are elements of the function spaces introduced in Chapter IV, Volume 2, or in Chapter I of the present Volume. The vector space S_α^β, for instance, is defined for $n = 1$, as follows

$$S_\alpha^\beta = \{\varphi\} = \{(\varphi_1, \dots, \varphi_m)\},$$

$$\| x^k \varphi^{(q)}(x) \| \leqslant C A^k B^q k^{k\alpha} q^{q\beta} \qquad (k, q = 0, 1, 2, \dots),$$

where $\| \ \|$ denotes the usual norm of a vector in Euclidean space. Such a space of vector test functions is simply a direct sum of identical scalar test function spaces.

The linear continuous functionals on such vector test function spaces can also be considered as vectors, "vectorial generalized functions" (or "vector-valued distributions"), $f = (f_1, \dots, f_n)$. The functional f acts on the vector test function $\varphi = (\varphi_1, \dots, \varphi_m)$ according to the equation

$$(f, \varphi) = (f_1, \varphi_1) + \cdots + (f_m, \varphi_m).$$

A linear operator A in the vector test function space Φ is defined by the matrix $\| A_{jk} \|$, where the elements A_{jk} are ordinary differential operators in the scalar test function space. The action of the operator A on the vector test function $\varphi = (\varphi_1, \dots, \varphi_m)$ is given by the usual rule

$$A\varphi = \begin{Vmatrix} A_{11} & \cdots & A_{1m} \\ \cdot & \cdots & \cdot \\ \cdot & \cdots & \cdot \\ \cdot & \cdots & \cdot \\ A_{m1} & \cdots & A_{mm} \end{Vmatrix} \cdot \begin{Vmatrix} \varphi_m \\ \cdot \\ \cdot \\ \cdot \\ \varphi_m \end{Vmatrix} = \begin{Vmatrix} \sum A_{1k}\varphi_k \\ \cdot \\ \cdot \\ \cdot \\ \sum A_{mk}\varphi_k \end{Vmatrix}.$$

The operator d/dx, in particular, is represented by a diagonal matrix, with diagonal elements equal to the ordinary differential operators d/dx. The operator of multiplication by x is represented by the diagonal matrix with diagonal elements x.

The adjoint of the operator A represented by the matrix $\| A_{jk} \|$ is the operator A^* defined by the matrix $\| A_{kj}^* \|$, which is obtained from $\| A_{jk} \|$ by replacing the matrix elements by their adjoints and transposing the matrix. For example, for $A = \| P_{jk}(id/dx) \|$, we have (id/dx is self-adjoint)

$$A^* = \left\| \bar{P}_{kj} \left(i \frac{d}{dx} \right) \right\|.$$

For simplicity we shall call a "vector function" a "function," and shall as a rule use a "scalar" terminology. It should be clear from the context whether the scalar or vector case is really meant. In the beginning we recommend to the reader to keep in mind only the scalar case, i.e., the case of a single equation. Some parts of the proofs which refer specifically to the case of systems of equations are omitted, but the reader should have no difficulty in reconstructing these parts.

According to the general plan outlined in Section 2, we consider the unknown function $u(x, t)$ as a generalized function, i.e., a linear functional, depending on the parameter t, over a vector test function space Φ. The problem reduces to the construction of such a test function space $\Phi = \{\varphi(x)\}$ in which the solution of the Cauchy problem for the system (vector notation!)

$$\frac{\partial \varphi(x, t)}{\partial t} = \tilde{P}\left(i \frac{\partial}{\partial x}\right) \varphi(x, t) \tag{1}$$

exists for arbitrary Cauchy data (initial conditions)

$$\varphi(x, t_0) = \varphi_0(x) \in \Phi. \tag{2}$$

Here the matrix $\tilde{P}$ is related to the matrix P by means of

$$\tilde{P}_{jk} = -\bar{P}_{kj} \qquad (j, k = 1, 2, \ldots, m).$$

The solution of the Cauchy problem can formally be written as

$$\varphi(x, t) = \exp\left[(t - t_0)\tilde{P}\left(i \frac{\partial}{\partial x}\right)\right] \varphi_0(x).$$

The matrix

$$\exp\left[(t - t_0)\tilde{P}\left(i \frac{\partial}{\partial x}\right)\right] = Q_{t_0}^{t}\left(i \frac{\partial}{\partial x}\right) = \left\| Q_{jk}\left(i \frac{\partial}{\partial x}, t_0, t\right) \right\|$$

has the elements $Q_{jk}(i(\partial/\partial x), t_0, t)$, which are entire analytic functions of the arguments $\partial/\partial x_1, \ldots, \partial/\partial x_n$ (i.e., $Q_{jk}(s_1, \ldots, s_n, t_0, t)$ is an entire analytic function of the complex variables $s_1, \ldots, s_n$). Thus the space Φ must be so constructed that the corresponding differential operators of infinite order be meaningful.

We know from Chapter IV, Volume 2 that the spaces $S_{\alpha,A}^{\beta,B}$ are suitable for this purpose, provided the parameters are adapted to the growth requirements of the functions $Q_{jk}(s, t_0, t)$. Therefore, we must first estimate the growth of the functions $Q_{jk}(s, t_0, t)$.

The norm of a matrix $A = \| a_{jk} \|$ $(j, k = 1, 2, \ldots, m)$ is defined as the norm of the corresponding linear operator in the m-dimensional

Euclidean space of vectors $\xi = (\xi_1, \ldots, \xi_m)$ with the scalar product $(\xi^{(1)}, \xi^{(2)}) = \sum \xi_j^{(1)}\xi_j^{(2)}$. It follows then[7]

$$\max_k \sum_{j=1}^{m} |a_{jk}|^2 \leqslant \| A \|^2 \leqslant \sum_{j=1}^{m} \sum_{k=1}^{m} |a_{jk}|^2. \tag{3}$$

Replacing A by the matrix $\tilde{P}(s)$ (the matrix of the system (1), with $i(\partial/\partial x)$ replaced by s), for sufficiently large $|s|$ one obtains for the matrix elements the estimate

$$|P_{jk}(s)| \leqslant C |s|^p,$$

since all matrix elements are polynomials of degree smaller than p in the variable s. The inequality (3) implies

$$\| \tilde{P}(s) \|^2 \leqslant \sum_{j=1}^{m} \sum_{k=1}^{m} |\tilde{P}_{jk}(s)|^2 \leqslant C_1^2 |s|^{2p}.$$

Further,

$$\| Q(s, t_0, t) \| = \| e^{(t-t_0)\tilde{P}(s)} \| = \| \sum_{k=0}^{\infty} \frac{(t-t_0)k}{k!} \tilde{P}^k(x) \|$$

$$\leqslant \sum_{k=0}^{\infty} \frac{(t-t_0)^k}{k!} \| \tilde{P}^k(s) \| \leqslant \sum_{k=0}^{\infty} \frac{(t-t_0)^k}{k!} \| \tilde{P}(s) \|^k$$

$$\leqslant \sum_{k=0}^{\infty} \frac{(t-t_0)^k}{k!} C_1^k |s|^{pk} \leqslant \exp[(t-t_0)C_1 |s|^p].$$

The elements of the matrix $Q(s, t_0, t)$ are obviously analytic functions of s. The last estimate shows that these matrix elements are actually *entire analytic functions of orders smaller or equal to p.*

Actually these functions may have a smaller order. We will show in Section 6 that there exists a minimal number $p_0 \leqslant p$ (the method of computing this number will be outlined there), such that the estimate

$$\| Q(s, t_0, t \| \leqslant C_2(1 + |s|)^{(m-1)p} \exp[b |t - t_0| |s|^{p_0}] \tag{4}$$

holds. The number p_0 is the exact exponential order of increase of the matrix and will be called the *reduced order of the system.*

[7] Cf., e.g., G. E. Shilov, Vvedenie v teoriyu lineinykh prostranstv (Introduction to the Theory of Linear Spaces), 2nd Russian Ed., p. 149, Moscow, 1956 (or any other standard text on the subject).

It should be clear for the reader from the context when $\| \; \|$ denotes a matrix and when the same symbol denotes the norm of the matrix.

The reduced order of a system characterizes its properties much more precisely than the order p (the highest of the orders of the differential operators with respect to x, occurring in the right-hand side of the system). The usual order can change when one replaces the unknown functions $u_j(x, t)$ by linear combinations of the unknown functions and their derivatives, whereas the reduced order does not change under such an operation. Consider, for example, the two systems:

$$\frac{\partial u_1}{\partial t} = u_2 , \quad \frac{\partial u_2}{\partial t} = a \frac{\partial^2 u_1}{\partial x^2} , \qquad \text{(I)} \qquad \frac{\partial v_1}{\partial t} = \frac{\partial v_2}{\partial x} , \quad \frac{\partial v_2}{\partial t} = a \frac{\partial v_1}{\partial x} , \qquad \text{(II)}$$

which transform into each other under the substitution $u_1 = v_1$, $u_2 = \partial v_2/\partial x$. As a matter of fact, both systems are different forms of writing the same second-order equation $u_{tt} = au_{xx}$. The usual orders of these two systems are nevertheless different (the first system has $p = 2$, the second has $p = 1$). It is easy to understand that the reduced orders of both systems are $p_0 = 1$. In a certain sense, the reduced order is the true order, and as has been shown by V. M. Borok, any system of reduced order p_0 can be transformed into a system for which the usual order is also p_0.

It should be noted that mutually inverse-adjoint systems having the matrices P and $\tilde{P} = -\bar{P}$ have the same reduced orders.

It happens that the reduced order of a system ultimately determines the uniqueness class for the Cauchy problem of a given system, according to the following fundamental theorem.

Theorem 1. *If a system of the form* (1), *Section* 3.1, *has the reduced order* $p_0 > 1$, *then the totality of functions* $f(x)$ *satisfying the inequalities*

$$|f(x)| \leqslant C \exp[b_0 x p_0'] \qquad \frac{1}{p_0} + \frac{1}{p_0'} = 1, \tag{5}$$

with arbitrary, but fixed C *and* b_0, *forms a uniqueness class of the Cauchy problem for the system* (1), *Section* 3.1. *In other words, there exists at most one solution of that system, which for* $t = 0$ *equals a given vector function* $u_0(x)$, *and all the components of which belong to the class* (5) *for fixed* t $(0 \leqslant t \leqslant T)$. *If* $p_0 = 1$, *the uniqueness class consists of the totality of functions* (5) *with arbitrary (fixed)* p_0'. *For* $p_0 < 1$ *the solution of the Cauchy problem for the systems* (1) *and* (2), *Section* 3.1 *is unique in the class of all functions* $f(x)$, *without any restrictions on their growth at infinity.*

The admissible interval of the parameter (time) $0 \leqslant t \leqslant T$ *depends only on the constants* C *and* b_0 *but does not depend on the choice of the initial function* $u_0(x)$.

The rest of this section is devoted to the proof of this theorem.

3.3. Proof of the Fundamental Theorem

We assume that t varies in the interval $t_0 \leqslant t \leqslant t_0 + T$, such that $bT < \theta$, where θ is arbitrary and b is the constant occurring in the estimate (4) of the preceding subsection. It follows from that estimate that the entire functions which make up the matrix $Q(s, t_0, t)$ are of order p_0 and type smaller than θ (of that order) (cf. Section 1, Chapter II, Volume 2).

We shall now make more precise the choice of the test function space Φ.

For simplicity let us assume first that the argument x is not an n-dimensional vector, but varies on the line $-\infty < x < \infty$.

The (scalar) space $S_{\alpha,A}^{\beta,B}$ consists of infinitely differentiable functions $\varphi(x)$ which for arbitrary $\delta > 0$, $\rho > 0$ satisfy the inequalities

$$| x^k \varphi^{(q)}(x) | \leqslant C_{\delta\rho}(A + \delta)^k k^{k\alpha}(B + \rho)^q q^{q\beta}.$$

In Section 5, Chapter IV, Volume 2, the following theorem is proved: if $f(s)$ is an entire function of order $1/\beta$ and type smaller than $\beta/(B^{1/\beta}e^2)$, then the operator $f(d/dx)$ is defined and bounded in the space $S_{\alpha,A}^{\beta,B}$ and maps this space into the space $S_{\alpha,A}^{\beta,B'}$, with $B' = Be^{\beta}$.

Since we now have estimates of the order (p_0) and type ($< \theta$) of the entire functions $Q_{ij}(s, t_0, t)$ we can make use of this theorem for the construction of the vector space $S_{\alpha,A}^{\beta,B}$ in which the operator $Q(i(\partial/\partial x), t_0, t)$ is defined. The numbers β and B are determined from the equations

$$\frac{1}{\beta} = p_0, \qquad \frac{\beta}{B^{1/\beta}e^2} = \theta. \tag{1}$$

The numbers α and A are determined from the condition that the space $S_{\alpha,A}^{\beta,B}$ be nontrivial. It follows from Eqs. (1) that:

$$\beta = 1/p_0, \qquad B = (\beta/\theta e^2)^{\beta}.$$

We first assume that $p_0 > 1$. Then $\beta < 1$ and in order to guarantee the nontrivial character of the space $S_{\alpha,A}^{\beta,B}$ we must assume (according to the results of Section 8, Chapter IV, Volume 2)

$$\alpha = 1 - \beta, \qquad A = \frac{\gamma}{B},$$

where γ is a positive number.

The case $p_0 \leqslant 1$ will be discussed somewhat later.

The test functions which make up the (scalar) space $S_{\alpha,A}^{\beta,B}$ are characterized by the following decrease at $|x| \to \infty$:

$$|\varphi(x)| \leqslant C \exp[-a|x|^{1/\alpha}].$$

The exponent $1/\alpha$ can be expressed in terms of the reduced order p_0

$$\frac{1}{\alpha} = \frac{1}{1-\beta} = \frac{1}{1-(1/p_0)} = p_0'.$$

The coefficient a can be computed according to Eq. (5) in Section 2.1, Chapter IV, Volume 2:

$$a = \frac{\alpha}{eA^{1/\alpha}};$$

it can be expressed in terms of the quantity θ:

$$a = \frac{\alpha}{e}\left(\frac{B}{\gamma}\right)^{1/\alpha} = C\theta^{-\beta/\alpha}. \tag{2}$$

Thus we have constructed the space $\Phi = S_{\alpha,A}^{\beta,B}$ in which the operator $Q(i(\partial/\partial x), t_0, t)$ acts. We now show that for any test function $\varphi(x) \in S_{\alpha,A}^{\beta,B}$ the solution of the Cauchy problem

$$\frac{\partial\varphi(x,t)}{\partial t} = \tilde{P}\left(i\frac{\partial}{\partial x}\right)\varphi(x,t), \qquad \varphi(x,t_0) = \varphi(x),$$

is given by the equation

$$\varphi(x,t) = Q\left(i\frac{\partial}{\partial x}, t_0, t\right)\varphi(x).$$

In other words, we have to show that for any test function $\varphi(x)$, we have in the topology of the space $S_{\alpha,A}^{\beta,Be^\beta}$

$$\lim_{\Delta t \to 0} \frac{\Delta\left[Q\left(i\frac{\partial}{\partial x}, t_0, t\right)\varphi(x)\right]}{\Delta t} = \tilde{P}\left(i\frac{\partial}{\partial x}\right)\left[Q\left(i\frac{\partial}{\partial x}, t_0, t\right)\varphi(x)\right],$$

$$\lim_{t \to t_0} Q\left(i\frac{\partial}{\partial x}, t_0, t\right)\varphi(x) = \varphi(x).$$

As we remember from Chapter IV, Volume 2, a sequence $\varphi_\nu(x)$ converges to a function $\varphi(x)$ in $S_{\alpha,A}^{\beta,Be^\beta}$ if the set of functions $\{\varphi_\nu(x)\}$ is bounded in this space and converges regularly in it to $\varphi(x)$, i.e., on each finite interval, the functions $\varphi_\nu^{(q)}(x)$ converge uniformly to $\varphi^{(q)}(x)$.

In order to apply this convergence test we note first that the matrices consisting of entire functions

$$Q(s, t_0, t) = e^{(t-t_0)\tilde{P}(s)}, \qquad \frac{\Delta Q(s, t_0, t)}{\Delta t} = \frac{e^{\Delta t P(s)} - 1}{\Delta t} e^{(t-t_0)\tilde{P}(s)},$$

admit the majorizations

$$\| Q(s, t_0, t) \| \leqslant \exp[\theta \mid s \mid^p], \qquad \left\| \frac{\Delta Q(s, t_0, t)}{\Delta t} \right\| \leqslant \exp[\theta \mid s \mid^p],$$

which do not depend on t, $0 \leqslant t \leqslant T$, $bT < \theta$, and thus define operators in the spaces $S_{\alpha,A}^{\beta,B}$ which are uniformly bounded (uniformity in t). It then follows that the families of functions

$$\psi(x, t) = Q(i(\partial/\partial x), t_0, t)\, \varphi(x),$$
$$\chi(x, t) = \frac{\Delta Q(i(\partial/\partial x), t_0, t)}{\Delta t}\, \varphi(x)$$

are bounded for $0 \leqslant t \leqslant T$ in the space $\Phi_1 = S_{\alpha,A}^{\beta,B'}$ $(B' = Be^{\beta})$. We have to show that the functions $\psi(x, t)$ converge regularly to $\varphi(x)$ as $t \to t_0$, and the functions $\chi(x, t)$ converge to $\tilde{P}(i(\partial/\partial x))\, \psi(x, t)$ as $\Delta t \to 0$.

The function $\psi(x, t)$ is defined by the series

$$\psi(x, t) = Q\left(i \frac{\partial}{\partial x}, t_0, t\right) \varphi(x) = \exp\left[(t - t_0)\tilde{P}\left(i \frac{\partial}{\partial x}\right)\right] \varphi(x)$$
$$= \sum_{k=0}^{\infty} \frac{(t - t_0)^k}{k!} \left[\tilde{P}^k \left(i \frac{\partial}{\partial x}\right) \varphi(x)\right],$$

which converges absolutely and uniformly in x, as has been proved for the scalar case in Section 5, Chapter I, Volume 2. The series converges uniformly also with respect to the parameter t within the limits $t_0 \leqslant t \leqslant T$. Therefore, one can take the limit as $t \to t_0$ term by term and sum up the results, obtaining

$$\lim_{t \to t_0} \psi(x, t) = \tilde{P}\left(i \frac{\partial}{\partial x}\right) \varphi(x) = \varphi(x).$$

Obviously, the series which are obtained by means of term-by-term differentiation with respect to x are also uniformly convergent. Thus the desired regular convergence is established.

The second limit relation is proved in the same manner.

Thus we have shown that $Q(i(\partial/\partial x), t_0, t)$ is the operator which solves the Cauchy problem in the space $\Phi = S_{\alpha,A}^{\beta,B}$.

We shall now make use of the theorem proved in Section 2.2, which asserts that if the operator $Q_{t_0}^t$ maps the space Φ linearly into the space $\Phi_1 \supset \Phi$, and for any $\varphi \in \Phi$, the function $\psi(t) = Q_{t_0}^t \varphi_0$ is a solution of the Cauchy problem

$$\frac{d\varphi(t)}{dt} = A\varphi(t), \qquad \varphi(t) = \varphi_0 ,$$

then the adjoint Cauchy problem

$$\frac{du(t)}{dt} = -A^* u(t), \qquad u(0) = u_0 ,$$

can have only a unique solution in the space E'. Here, E' is the dual of any space E which contains the spaces Φ and Φ_1 as dense subsets.

In the case under consideration, as stated earlier, the operators $Q_{ij}(i(\partial/\partial x), t_0, t)$ map the space $\Phi = S_{\alpha,A}^{\beta,B}$ with the indicated values of α, β, A, B into the space $\Phi_1 = S_{\alpha,A}^{\beta,Be^\beta}$. This space Φ_1 consists of functions $\varphi(x)$ which have the same character of decrease at infinity as the functions of the space Φ. Consequently this space is contained in the same normed space E, which consists of measurable functions $\varphi(x)$ $(-\infty < x < \infty)$ with the norm defined by

$$\| \varphi \|_E = \int_{-\infty}^{\infty} \exp[\tfrac{1}{2}a \mid x \mid^{p_0'}] \mid \varphi(x) \mid dx. \tag{3}$$

According to the results of Section 8, Chapter I, Volume 2, the nontrivial space $S_{\alpha,A}^{\beta,B}$ is sufficiently rich in functions and is densely contained in any larger normed space E with a norm of type (3). Thus the hypotheses of the theorem in Section 2.2. are verified and we reach the conclusion: The Cauchy problem for the system (1) in Section 3.1 can have only one solution belonging to the space E' for all t, $0 \leqslant t \leqslant T$, $bT < \theta$.

The dual E' of the space E consists of all measurable functions $f(x)$ which almost everywhere satisfy the inequality

$$\| f(x) \| < C \exp[\tfrac{1}{2}a \mid x \mid^{p_0'}]. \tag{4}$$

Consequently *the Cauchy problem* (1) *and* (2) *of Section* 3.1 *has a unique solution in the class of measurable functions satisfying the inequality* (4) *for* $0 \leqslant t \leqslant T$, $bT < \theta$.

Note that due to the narrowing of the admissible interval of the time variable $0 \leqslant t \leqslant T$, $bT < \theta$, the constant $a/2$ in the inequality (4), which is related to θ via Eq. (2), can be arbitrarily large. In particular, we can choose this constant equal to a given b_0.

Thus, Theorem 1 has been proved for $p_0 > 1$.

Let us now consider the remaining cases $p_0 = 1$ and $p_0 < 1$.

For $p_0 = 1$, the number β determined by Eq. (1) equals 1 also, so that one may no longer choose $\alpha = 1 - \beta = 0$, since the corresponding space S_0^1 is trivial. Therefore we choose an arbitrary positive number for α. According to Section 8, Chapter IV, Volume 2, the space $S_{\alpha,A}^{1,B}$ is nontrivial, and it can be taken as the space Φ used in the proof of the theorem.

We thus arrive at the uniqueness class of the Cauchy problem (1) and (2), Section 3.1, characterized by

$$|f(x)| \leqslant \exp[b_0 \, |x|^{1/\alpha}],$$

where $1/\alpha$ is an arbitrary but fixed positive number.

Finally, for $p_0 < 1$, we have $\beta = 1/p_0 > 1$, and we can choose $\alpha = 0$. Although the space $S_{0,A}^{\beta,B}$ is nontrivial, it is not sufficiently rich in test functions. Therefore we choose instead the space $S_0^{\beta,B}$ as the space Φ. The space $S_0^{\beta,B}$ is the union with respect to the subscript A of all spaces $S_{0,A}^{\beta,B}$. This space is already sufficiently rich in functions.

All functions belonging to the space $\Phi = S_0^{\beta,B}$ have compact support. Thus the space Φ is included in the Banach space E with norms defined as

$$\| \varphi \|_E = \int_{-\infty}^{\infty} E(x) \, \| \varphi(x) \| \, dx$$

where $E(x)$ is an *arbitrary* weight function.

We thus obtain the result that the class of functions $f(x)$ satisfying the inequality

$$|f(x)| \leqslant CE(x),$$

is a uniqueness class of solutions of the Cauchy problem, Eqs. (1) and (2), in Section 3.1. Since $E(x)$ is an arbitrary function, the solution is unique within the class of all functions $f(x)$, without any restrictions on their growth. Thus for $n = 1$ we have completed the proof of the theorem.

For $n > 1$ independent variables, $x_1, \ldots, x_n$ (not a single x, as before), the proof can be carried out according to the same plan, with the space $S_{\alpha,A}^{\beta,B}$ replaced by its n-dimensional counterpart $S_{\alpha_1,\ldots,\alpha_n,A_1,\ldots,A_n}^{\beta_1,\ldots,\beta_n,B_1,\ldots,B_n}$ (cf. Section 9, Chapter IV, Volume 2). The functions $Q_{jk}(s, t_0, t) = Q_{jk}(s_1, \ldots, s_n, t_0, t)$ now satisfy the inequalities

$$|Q_{jk}(s_1, \ldots, s_n, t_0, t)| \leqslant C \exp[\theta_1 \, |s_1|^{p_0} + \cdots + \theta_n \, |s_n|^{p_0}] \tag{5}$$

with reduced order p_0. As before, the numbers β_j, B_j, α_j, A_j are chosen according to the value of p_0. The ensemble of these numbers determines the space $S^{\beta_1,\ldots,\beta_n,B_1,\ldots,B_n}_{\alpha_1,\ldots,\alpha_n,A_1,\ldots,A_n}$. As a result, we arrive at a uniqueness class characterized by the inequalities

$$|f(x_1,\ldots,x_n)| \leqslant C \exp[a_1 |x_1|^{p_0'} + \cdots + a_n |x_n|^{p_0'}] \tag{6}$$

for $p_0 > 1$. For $p_0 = 1$, we obtain similar inequalities, but with p_0' replaced by an arbitrary power r. Finally, for $p_0 < 1$, we arrive at the class of all functions, without any restrictions on their growth. This completes the proof of Theorem 1 for arbitrary dimensions.

We note a new circumstance which occurs in the case of several variables. In addition to the majorization (5), the entire functions $Q_{ij}(s_1,\ldots,s_n, t_0, t)$ admit also the estimate

$$|Q_{jk}(s_1,\ldots,s_n, t_0, t)| \leqslant C \exp[\theta_1 |s_1|^{p_1^0} + \cdots + \theta_n |s_n|^{p_n^0}],$$

where the numbers p_j^0 may be either smaller or larger than the reduced order p_0. Selecting the numbers $\beta_j, B_j, \alpha_j, A_j$, which determine the space $S^{\beta_1,\ldots,\beta_n,B_1,\ldots,B_n}_{\alpha_1,\ldots,\alpha_n,A_1,\ldots,A_n}$ in agreement with these values p_j^0, we obtain a uniqueness class characterized by the inequality

$$|f(x_1,\ldots,x_n)| \leqslant C \exp[a_1 |x_1|^{q_1^0} + \cdots + a_n |x_n|^{q_n^0}] \qquad \left(\frac{1}{p_j^0} + \frac{1}{q_j^0} = 1\right).$$

In general, this class is not contained in the class defined by the inequality (6). Taking for example the function

$$Q(s_1, s_2, t_0, t) = e^{(t-t_0)s_1s_2}$$

we can write the inequality

$$|\exp[(t-t_0)s_1s_2]| \leqslant \exp[(t-t_0)|s_1||s_2|] \leqslant \exp\left[(t-t_0)\left(\frac{|s_1|^{r_1}}{r_1} + \frac{|s_2|^{r_2}}{r_2}\right)\right],$$

with $(1/r_1) + (1/r_2) = 1$, but r_1 and r_2 otherwise arbitrary. Accordingly, for the equation

$$\frac{\partial u}{\partial t} = \frac{\partial^2 u}{\partial x_1 \partial x_2}$$

any of the classes of functions characterized by the inequalities

$$|f(x_1, x_2) \leqslant C \exp[b_1 |s_1|^{r_1} + b_2 |s_2|^{r_2}] \qquad \left(\frac{1}{r_1} + \frac{1}{r_2} = 1\right)$$

serves as a uniqueness class.

3.4. Ordinary Solutions as Generalized Solutions

We must complete our analysis in one important respect. Until now the unknown $u_j(x, t)$ $(j = 1,..., m)$ in the Cauchy problem

$$\frac{\partial u_j(x, t)}{\partial t} = \sum_{k=1}^{m} P_{jk}\left(i\frac{\partial}{\partial x}\right) u_k(x, t), \tag{1}$$

$$u_j(x, 0) = u_j(x) \tag{2}$$

were considered as generalized functions over a test function space Φ. Likewise, the operators $\partial/\partial t$ and $P_{jk}(i(\partial/\partial x))$ which occur in Eq. (1) were understood in the sense of the theory of generalized functions: in order words, the fact that the system (1) is verified for certain $u_j(x, t)$ means that for any test function $\varphi(x)$ we have in effect

$$\frac{\partial}{\partial t}(u_j(x, t), \varphi(x)) = \sum_{k=1}^{m}\left(u_k(x, t), \tilde{P}_{jk}\left(i\frac{\partial}{\partial x}\right)\varphi(x)\right),$$

and the initial conditions (2) mean actually

$$\lim_{t\to 0}(u_j(x, t), \varphi(x)) = (u_j(x, 0), \varphi(x)).$$

(i.e., we are dealing with the problem "weakly").

Therefore, before applying Theorem 1 to an *ordinary* solution $u_j(x, t)$ of the Cauchy problem, we must check that the corresponding generalized function

$$(u_j(x, t), \varphi(x)) = \int u_j(x, t)\varphi(x)\, dx$$

is a solution of the problem (1) and (2) in the sense of the theory of generalized functions.

This is guaranteed under quite general conditions by the following

Theorem 2. *Let $u_k(x, t)$ $(k = 1, 2,..., m)$ be ordinary functions which are differentiable with respect to t and which admit the application of the differential operators $P(i(\partial/\partial x))$ (i.e., which admit derivatives with respect to x up to order p). Let these functions $u_k(x, t)$ transform the system (1) into a system of identities and satisfy the inequalities*

$$|u_k(x, t)| \leqslant C\exp[\tfrac{1}{2}a\,|x|^{1/\alpha}] \tag{3}$$

Then the system of functionals

$$(u_k(x, t), \varphi(x)) = \int_{-\infty}^{\infty} u_k(x, t)\varphi(x)\, dx \tag{4}$$

defined on the space $\Phi = S_{\alpha,A}^{\beta,B}$, $a = \alpha/eA^{1/\alpha}$, *is a solution of the Cauchy problem for the system* (1) *in the sense of generalized functions. The initial functionals are determined by the functions* $u_k(x, 0)$.

Proof. We note first that due to condition (3), the expressions (1) define indeed linear continuous functionals on the space $S_{\alpha,A}^{\beta,B}$. Moreover, due to the t-independence of the constants in (3), the expressions (4) are uniformly continuous with respect to t in the following sense: if a sequence of functions $\varphi_\nu(x)$ converges to $\varphi(x)$ in the topology of the space $S_{\alpha,A}^{\beta,B}$, then the limit relation

$$(u_k(x, t), \varphi_\nu(x)) \to (u_k(x, t), \varphi(x))$$

holds uniformly with respect to t in the interval $0 \leqslant t \leqslant T$.

It remains to be shown that for each test function $\varphi(x)$ we have the equalities

$$\frac{\partial}{\partial t} \int u_k(x, t)\varphi(x)\, dx = \sum \int u_j(x, t)\bar{P}_{jk}\left(i\frac{\partial}{\partial x}\right) \varphi(x)\, dx \tag{5}$$

and the limit relations

$$\int [u_k(x, t) - u_k(x, 0)]\varphi(x)\, dx \to 0. \tag{6}$$

We cannot simply multiply the system by $\varphi(x)$ and integrate by parts, since nothing guarantees the convergence of the integrals so obtained. This necessitates the use of a roundabout way of reasoning.

The functional $u_k(x, t)$ can be extended from the spaces $S_{\alpha,A}^{\beta,B}$ to the space $S_{\alpha,A}$. The latter consists of infinitely differentiable functions $\varphi(x)$ satisfying the inequalities

$$|\varphi^{(q)}(x)| \leqslant C_q \exp[-a_1 |x|^{1/\alpha}]$$

where $a_1 = \alpha/(A + \delta)^{1/\alpha}e$.

Let $\varphi(x) \in S_{\alpha,A}$. As was shown in Volume 2, functions of compact support are a dense subset of this space, so that one can construct a sequence of functions of compact support $\{\varphi_\nu(x)\}$ which converges to the function $\varphi(x)$ in the topology of the space $S_{\alpha,A}$. The continuity of the functional $u_k(x, t)$ then implies

$$(u_k(x, t), \varphi_\nu(x)) \to (u(x, t), \varphi(x)) \tag{7}$$

uniformly in t $(0 \leqslant t \leqslant T)$.

The functions $\psi_\nu(x) = \bar{P}_{jk}(i(\partial/\partial x))\, \varphi_\nu(x)$ are also of compact support, and due to the continuity of the differentiation operators in $S_{\alpha,A}$ we obtain

$$\begin{gathered}\bar{P}_{jk}\left(i\frac{\partial}{\partial x}\right)\varphi_\nu(x) \to \bar{P}_{jk}\left(i\frac{\partial}{\partial x}\right)\varphi(x),\\ \left(u(x, t), \bar{P}_{jk}\left(i\frac{\partial}{\partial x}\right)\varphi_\nu(x)\right) \to \left(u(x, t), \bar{P}_{jk}\left(i\frac{\partial}{\partial x}\right)\varphi(x)\right),\end{gathered} \tag{8}$$

the latter being uniform in t.

Multiplying the equality

$$\frac{\partial u_k(x, t)}{\partial t} = \sum P_{jk}\left(i\frac{\partial}{\partial x}\right) u_j(x, t).$$

with the function of compact support $\varphi_\nu(x)$, and integrating by parts, we obtain:

$$\frac{\partial}{\partial t}\int u_k(x, t)\varphi_\nu(x)\, dx = \sum_j \int u_j(x, t)\bar{P}_{jk}\left(i\frac{\partial}{\partial x}\right)\varphi_\nu(x)\, dx. \tag{9}$$

This, together with (8), implies that the functions appearing in the left-hand side of the equality (9) have a limit as $\nu \to \infty$ which is equal to the right-hand side of (8), and converge uniformly to this limit. But according to (7), the function $(u(x, t), \varphi_\nu(x))$ themselves converge to $(u(x, t), \varphi(x))$ as $\nu \to \infty$. Then the theorem on differentiation of functional sequences implies

$$\frac{\partial}{\partial t}(u_k(x, t), \varphi(x)) = \left(u_k(x, t), \sum \bar{P}_{jk}\left(i\frac{\partial}{\partial x}\right)\varphi(x)\right),$$

i.e., the equality (5) that we were required to prove.

In order to prove (6) we note that this equality is obviously true for functions of compact support $\varphi(x) \in S_{\alpha,A}$. Since the functionals $u_k(x, t) - u_k(x, 0)$ are uniformly (in t) bounded (due to the inequalities (3)) and converge to zero on a set of test functions of compact support, they also converge to zero on any element $\varphi \in S_{\alpha,A}$. This accomplishes the proof of Theorem 2.

4. The Cauchy Problem for Systems of Partial Differential Equations. The Method of Fourier Transforms

4.1. Introduction

In this section we shall rederive the results of Section 3, making use of a proof based on Fourier transforms.

Already the "natural philosophers" of the nineteenth century made use of the Fourier transform for solving the Cauchy problem. However, up to very recently, the applicability of this method was restricted by various requirements on the behavior of the transformed functions at infinity. In the classical papers on the subject, conditions like the integrability of the functions or of certain powers of the functions were usually assumed. In our century, such requirements were replaced by conditions of integrability of the product of the function with an exponential $e^{-\lambda x}$ (a Laplace transform, which is one of the forms of Fourier transforms). One could not push the methods of classical mathematics any further, and therefore, one made use of other methods in the investigation of partial differential equations, as for instance, the method of characteristics for hyperbolic equations. The method of characteristics is quite natural for hyperbolic equations. It leads to solutions, no matter what the growth characteristics of the initial data are at infinity. At the same time, the classical form of the method of Fourier transforms yields solutions only if the initial data do not increase too fast.

Since we utilize generalized functions we have the possibility of Fourier-transforming any arbitrarily rapidly increasing functions. This allows, in a manner of speaking, to "rehabilitate" the method of Fourier transforms in the treatment of the Cauchy problem. This leads to a uniform construction of the solution for all types of systems (and not only for hyperbolic systems, where the method of characteristics operates).

The method of Fourier transforms also gives an approach to the problem of correctness classes of solutions of the Cauchy problem. These problems will be treated in the next chapter.

4.2. The Fundamental Theorem

We consider the Cauchy problem for the system of partial differential equations (we use the vector notation)

$$\frac{\partial u(x, t)}{\partial t} = P\left(i \frac{\partial}{\partial x}\right) u(x, t), \tag{1}$$

with the initial data

$$u(x, 0) = u_0(x). \tag{2}$$

As in Section 3, we pose the problem of determining the uniqueness class of solutions of the Cauchy problem (1) and (2). We again consider the solution $u(x, t)$ a generalized (vector) function over a test function space, depending in a continuous and differentiable manner on the parameter t.

According to the fundamental result of Section 2, the Cauchy problem (1) and (2) will have a unique solution in the space Φ', if for any initial test function $\varphi_0(x)$, there exists a solution for the adjoint Cauchy problem

$$\frac{\partial\varphi(x, t)}{\partial t} = \tilde{P}\left(i\frac{\partial}{\partial x}\right)\varphi(x, t) \tag{3}$$

$$\varphi(x, t_0) = \varphi_0(x), \tag{4}$$

(where $-\tilde{P}\ (i(\partial/\partial x))$ is the Hermitean adjoint of the operator $P(i(\partial/\partial x))$ (i.e., $\tilde{P}_{jk} = -\bar{P}_{kj}$).

In order to find a suitable space Φ, we Fourier-transform the problem (3) and (4). The system (3) transforms into the system of ordinary differential equations

$$\frac{d\psi(s, t)}{dt} = \tilde{P}(s)\psi(s, t), \tag{5}$$

where s is a parameter; the initial condition (4) is replaced by

$$\psi(s, t_0) = \psi_0(s), \tag{6}$$

where $\psi_0(s)$, the Fourier transform of the test function $\varphi_0(x)$, belongs to the test function space $\Psi = \tilde{\Phi}$.

Formally, the solution of the Cauchy problem (5) and (7) can be written as

$$\psi(s, t) = \exp[(t - t_0)\tilde{P}(s)]\, \psi_0(s).$$

The matrix $\exp((t - t_0)\, \tilde{P}(s)) = Q(s, t_0, t)$ consists of entire analytic functions of s. Therefore the space Ψ must be constructed in such a manner as to allow multiplication with such entire functions.

We first consider the case of a single independent variable; the transition to several variables is carried out in the same manner as in Section 3.

We make use of the spaces $W_{M,a}^{\Omega,b}$ constructed in Chapter I, choosing the parameters of these spaces according to the growth properties of the function $Q(s, t_0, t)$. These growth properties have been indicated in Section 3 and are characterized by the inequality

$$\| Q(s, t_0, t) \| \leqslant C_1(1 + | s |)^{(m-1)p_0} \exp[b_0(t - t_0) | s |^{p_0}],$$

where p_0 is the reduced order of the system. If t varies within the interval $t_0 \leqslant t \leqslant t_0 + T$, with $2^{p_0+1}b_0T < (1/p_0)\, \theta^{p_0}$, one can also write

$$\| Q(s, t_0, t) \| \leqslant C_2 \exp[2 - p_0\theta^{p_0} | s |^{p_0}(1/p_0)].$$

Since $|s|^{p_0} = |\sigma + i\tau|^{p_0} \leqslant 2^{p_0}(|\sigma|^{p_0} + |\tau|^{p_0})$, we have finally

$$\|Q(s, t_0, t)\| \leqslant C_2 \exp[(1/p_0)\theta^{p_0}(|\sigma|^{p_0} + |\tau|^{p_0})].$$

We recall that the space $W_{M,a}^{\Omega,b}$ consists of entire analytic functions $\psi(s)$ which satisfy the inequalities

$$\psi(\sigma + i\tau) \leqslant C \exp[-M(\bar{a}\sigma) + \Omega(\bar{b}\tau)],$$

where M and Ω are convex functions (with convexity downward); $\bar{a}$ is an arbitrary constant smaller than a and $\bar{b}$ is an arbitrary constant larger than b.

In Section 2, Chapter I, we have proved the following theorem:

Any entire analytic function $f(s)$ satisfying the inequality

$$|f(\sigma + i\tau)| \leqslant C \exp[M(a_1\sigma) + \Omega(b_1\tau)],$$

defines a (continuous and bounded) multiplication operator (multiplier) in the space $\Psi = W_{M,a}^{\Omega,b}$, for $a > a_1$. This operator maps the space $\Psi = W_{M,a}^{\Omega,b}$ into the space $\Psi_1 = W_{M,a-a_1}^{\Omega,b+b_1}$.

We assume that $p_0 > 1$.

We choose $M(\sigma) = (1/p_0)|\sigma|^{p_0}$, $\Omega(\tau) = (1/p_0)|\tau|^{p_0}$; these functions are convex and we have accordingly

$$\|Q(s, t_0, t)\| \leqslant C_2 \exp[M(\theta\sigma) + \Omega(\theta\tau)]$$

We now apply the above-mentioned theorem. For a given a, one can always choose the "time" interval $0 \leqslant t \leqslant T$ so small that the inequality $\theta < a$ holds; for such values of T the matrix $Q(s, t_0, t)$ will be a multiplier in the space $W_{M,a}^{\Omega,b}$ which maps this space into the space $W_{M,a-\theta}^{\Omega,b+\theta}$.

In order to show that for any test function $\psi_0(s) \in \Psi = W_{M,a}^{\Omega,b}$ the function

$$\psi(s, t) = Q(s, t_0, t)\psi_0(s)$$

solves the Cauchy problem

$$\frac{\partial\psi(s, t)}{\partial t} = \tilde{P}(s)\psi(s, t), \quad \psi(s, t_0) = \psi_0(s),$$

and belongs for $0 \leqslant t \leqslant T$ to the space $\Psi_1 = W_{M,a-\theta}^{\Omega,b+\theta}$, we must show that

$$\lim_{\Delta t \to 0} \frac{\Delta Q(s, t_0, t)\psi_0(s)}{\Delta t} = \tilde{P}(s)Q(s, t_0, t)\psi_0(s), \tag{7}$$

$$\lim_{t \to t_0} Q(s, t_0, t)\psi_0(s) = \psi_0(s). \tag{8}$$

in the topology of the space Ψ_1.

According to the definition of convergence in the space $W_{M,a}^{\Omega,b}$ (cf. Section 1, Chapter I) we must show that the functions under the lim sign converge regularly to their limits (i.e., they converge uniformly in each bounded domain of the s-plane) and are uniformly bounded with respect to t (in the sense of the space Ψ_1). The regular convergence is obvious in our case from the definition of the function $Q(s, t_0, t) = \exp((t - t_0)\tilde{P}(s))$. The uniform boundedness in t can be inferred from the inequalities

$$\| Q(s, t_0, t) \| \leqslant C_2 \exp[M(\theta\sigma) + \Omega(\theta\tau)]$$

and

$$\left\| \frac{\Delta Q(s, t_0, t)}{\Delta t} \right\| \leqslant C_3 \exp[M(\theta\sigma) + \Omega(\theta\tau)],$$

which do not depend on t.

Thus the limit relations (7) and (8) are true.

The isomorphism of the spaces Ψ and Φ implies that there exists an operator on the space Φ which solves the Cauchy problem (3) and (4) for any initial test function:

$$\frac{\partial \varphi(x, t)}{\partial t} = \tilde{P}\left(i \frac{\partial}{\partial x}\right) \varphi(x, t), \qquad \varphi(x, 0) = \varphi_0(x).$$

Therefore, according to Section 2, the solution of the Cauchy problem

$$\frac{\partial u(x, t)}{\partial t} = P\left(i \frac{\partial}{\partial x}\right) u(x, t), \qquad u(x, 0) = u_0(x),$$

is unique in the dual space of Φ.

More precisely, the solution $u(x, t)$ is unique, if for every t in the interval $0 \leqslant t \leqslant T$ it belongs to the space E', where E is a normed space containing the spaces Φ and Φ_1 as dense subspaces.

Since it is Fourier-dual to the space $\Psi = W_{M,a}^{\Omega,b}$, the space Φ is $W_{M_1,1/b}^{\Omega_1,1/a}$, where M_1 and Ω_1 are dual in the sense of Young to the functions Ω and M, respectively. Since $\Omega(\sigma) = M(\sigma) = (1/p_0)|\sigma|^{p_0}$, $M_1(x) = \Omega_1(x) = (1/p_0')|x|^{p_0'}$. According to the definition of the space $W_{M_1,1/b}^{\Omega_1,1/a}$ (Chapter I), the space Φ consists of functions $\varphi(x)$ which decrease along the x-axis as

$$\| \varphi(x) \| \leqslant C \exp\left[-\frac{1}{p_0'} \left[\frac{|x|}{b}\right]^{p_0'}\right].$$

Similarly the functions $\varphi(x)$ which form the space $\Psi_1 = \Phi_1 = W^{\Omega_1,(1/a-\theta)}_{M_1,1/(b+\theta)}$, decrease along the x-axis according to the inequality

$$\| \varphi(x) \| \leqslant C \exp \left[-\frac{1}{p_0'} \left[\frac{|x|}{b+\theta} \right]^{p_0'} \right]$$

We define the space E by means of the norm

$$\| \varphi \|_E = \int_{-\infty}^{\infty} \exp[|x|^{p_0'}/2p_0'b] \, \| \varphi(x) \| \, dx.$$

Obviously $\Phi \subset \Phi_1 \subset E$. Since Φ is sufficiently rich in functions (Section 1, Chapter I), it is dense in E. We can thus apply Theorem 2, Section 2 and arrive at the conclusion that *the Cauchy problem* (1) *and* (2) *has a unique solution in the space* E'.

This immediately leads to the fundamental theorem on uniqueness classes.

Theorem. *For sufficiently small* $t \leqslant T$, *the Cauchy problem* (1) *and* (2) *admits a unique solution in the class of functions* $f(x)$, *which for fixed arbitrary* b_0 *satisfy the inequality*

$$|f(x)| \leqslant C \exp[b_0 |x|^{p_0'}].$$

Indeed, for a given b_0, we can determine b from the condition $1/(2p_0'b) = b_0$. Then we determine a from the condition that the space $W^{\Omega,b}_{M,a}$ be nontrivial (Chapter I, Section 1) with $M(x) = \Omega(x) = (1/p_0')|x|^{p_0'}$, and finally we determine the interval T from the condition $2^{p_0+1}b_0T < (1/p_0)\,\theta^{p_0}(\theta < a)$. These quantities allow us to construct the test function spaces Φ and Ψ and to carry out the reasoning which achieves the proof of the theorem.

The uniqueness theorem has been proved for $p_0 > 1$. For $p_0 \leqslant 1$, the proof is carried out according to the same plan, but replacing the functions $|\sigma|^{p_0}$, $|\tau|^{p_0}$ which define the space $W^{\Omega,b}_{M,a}$, by the functions $|\sigma|^r$, $|\tau|^r$ with arbitrary positive powers $r > 1$. The uniqueness class so obtained will be characterized by the inequality

$$|f(x)| \leqslant C \exp[b_0 |x|^{r'}],$$

where r' can be considered an arbitrary fixed positive number.

Note. Our proof has made use of the space $W_M{}^\Omega$. It would have been possible to use different spaces too. The only essential restriction is that within the space under consideration the multiplication by functions $e^{tP(s)}$ should be a bounded multiplier.

4.3. The Case of Hyperbolic Systems

The Fourier method no longer yields for $p_0 < 1$ a uniqueness class consisting of all functions without restrictions on their growth (cf. Section 3). On the other hand, in one important case it allows to improve upon the result obtained in Section 2, for $p_0 = 1$.

Let us assume specifically that the matrix function $Q(s, t_0, t) = \exp[(t - t_0)\tilde{P}(s)]$ satisfies the inequalities

$$\begin{aligned} &\text{(a)} \quad \| Q(s, t_0, t) \| \leqslant C_1 e^{\theta |s|} && \text{for all } s \text{ (i.e., } p_0 \leqslant 1); \\ &\text{(b)} \quad \| Q(\sigma, t_0, t) \| \leqslant C_2(1 + |\sigma|^h), && \text{for real } s = \sigma, \end{aligned} \tag{1}$$

i.e., its growth is not faster than a power.

In the sequel, we shall call *hyperbolic* systems of the form (1), Section 4.2, for which the resolvent matrix $Q(s, t_0, t)$ posesses these properties. Such systems will be dealt with in more detail in Section 3, Chapter III.

We assert that *for hyperbolic systems* (as well as for systems with $p_0 < 1$) the Cauchy problem has a unique solution in the class of all functions, without restrictions on the growth at infinity.

For the proof we make use of Theorems 1′ and 2′ in Chapter IV, Section 7, Volume 2. We have proved there that each entire function which satisfies the inequalities (1) and (2), also satisfies the inequalities

$$\| Q(s, t_0, t) \| \leqslant C_3 e^{\theta|\tau|}(1 + |\sigma|^h).$$

Then, on the basis of Theorem 3″ in the same section of Volume 2, the function $Q(s, t_0, t)$ is a multiplier in the space $Z = S^0$ of entire analytic functions $\psi(s)$ which satisfy the inequalities

$$\| s^k \psi(s) \| \leqslant C_k e^{b|\tau|}.$$

It follows that the Fourier-transformed operator $G(x, t_0, t) = F[Q(s, t_0, t)]$ is defined on the space $K = S_0$ of infinitely differentiable functions of compact support and that this operator maps the space K into itself. Any (locally integrable) function, without restrictions on its growth, defines a functional belonging to the class of generalized functions over K. Therefore, repeating the above arguments we find that the uniqueness of the solution of the Cauchy problem is guaranteed within the class of all functions, as asserted.

4.4. Systems with Coefficients Depending on t

All the results obtained so far can be extended to systems with coefficients which depend on t continuously:

$$\frac{\partial u(x, t)}{\partial t} = P\left(i\frac{\partial}{\partial x}, t\right) u(x, t). \tag{1}$$

We sketch the proof using the method of Fourier transforms. The operator method could have been used with equal success.

It suffices to prove the existence of the solution of the Cauchy problem for the system in the test function space $\Phi = \{\varphi(x)\}$

$$\frac{\partial \varphi(x, t)}{\partial t} = \tilde{P}\left(i\frac{\partial}{\partial x}, t\right) \varphi(x, t).$$

As before $-\tilde{P}$ denotes the hermitean adjoint of the matrix P.

The Fourier-transformed Cauchy problem is

$$\frac{d\psi(s, t)}{dt} = \tilde{P}(s, t)\psi(s, t), \tag{2}$$

$$\psi(s, 0) = \psi_0(s), \tag{3}$$

where $\psi_0(s)$ is a test function in the space $\Psi = \tilde{\Phi}$. It is no longer possible to write the solution of this problem in the form of an exponential matrix function, as we did before. We have to use a different procedure.

On the basis of classical existence theorems for ordinary differential systems, the system (2) admits a normal fundamental matrix of solutions

$$Q(s, t_0, t) = \| Q_{jk}(s, t_0, t) \|,$$

corresponding to the initial conditions

$$Q(s, t_0, t_0) = E \text{ (the unit matrix)}.$$

For a given initial vector $\psi_0(s)$, the solution of the system (2) which for $t = t_0$ becomes equal to $\psi_0(s)$ has the form

$$\psi(s, t) = Q(s, t_0, t)\psi(s).$$

Since the system (2) is analytic in the parameters s, the matrix elements $Q_{ij}(s, t_0, t)$ are entire analytic functions, for which we can evaluate the order of growth.

The matrix $Q(s, t_0, t)$ can be constructed in the following manner.[8]

[8] Cf., e.g., F. R. Gantmakher, Teoriya matrits (Matrix Theory), Chap. 14, Gostekhizdat, Moscow, 1953. An English translation seems to be available.

We partition the interval (t_0, t) into n parts by means of the points $t_0 < t_1 < \cdots < t_n = t$ and set $\Delta t = t_i - t_{i-1}$ $(i = 1,\ldots, n)$. Making use of the composition property of the matrix $Q(s, t_0, t)$ (cf. Section 2), we can write

$$Q(s, t_0, t) = Q(s, t_{n-1}, t_n)Q(s, t_{n-2}, t_{n-1}) \cdots Q(s, t_0, t_1).$$

Keeping s fixed and making use of Eq. (7), Section 2.1, with $t_0 = t_i$, we can write

$$Q(s, t_i, t_{i+1}) = E + \tilde{P}(s, t_i)\,\Delta t_i + \epsilon(\Delta t_i),$$

where $\epsilon(\Delta t_i) \to 0$ as $\Delta t_i \to 0$, uniformly in the interval $0 \leqslant t \leqslant T$. Hence

$$Q(s, t_0, t) = \prod_{j=0}^{n-1} [E + \tilde{P}(s, t_i)\Delta t_i] + \epsilon(\max \Delta t_i).$$

One can now go to the limit $\max \Delta t_i \to 0$, obtaining

$$Q(s, t_0, t) = \lim_{max \Delta t_i - 0} \prod_{j=0}^{n-1} [E + \tilde{P}(s, t_i)\Delta t_i].$$

This expression allows one to evaluate the growth of the matrix $Q(s, t_0, t)$ with respect to s. Since the polynomials $\tilde{P}_{jk}(s, t_i)$ have degrees not larger than p, it follows that for sufficiently large $|s|$ (independent of $t \leqslant T$)

$$|\tilde{P}_{jk}(s, t_i)| \leqslant C' |s|^p,$$

whence

$$\|\tilde{P}(s, t_i)\| \leqslant C |s|^p.$$

Taking $\Delta t_i = (t - t_0)/n$, we obtain the estimate

$$\left\| \prod_{t_0}^{t} [E + \tilde{P}(s, t_i)\Delta t_i] \right\| \leqslant \left(1 + C |s|^p \frac{t - t_0}{n}\right)^n \leqslant \exp[C |s|^p(t - t_0)].$$

In order to obtain an estimate which holds for all values of s it is sufficient to introduce an additional factor C_1, so that for all s

$$\left\| \prod_{t_0}^{t} [E + \tilde{P}(s, t_i)\,\Delta t_i] \right\| \leqslant C_1 \exp[C |s|^p(t - t_0)].$$

Since this estimate does not depend on the partition of the interval (t_0, t) one can go to the limit letting max Δt_i approach zero. This yields

$$\| Q(s, t_0, t) \| \leqslant C_1 \exp[C | s |^p (t - t_0)].$$

Thus the entire analytic functions of s making up the matrix $Q(s, t_0, t)$ are of order not larger than p and type $\leqslant C(t - t_0)$.

In this case also, the order of growth of the matrix $Q(s, t_0, t)$ may be lower than p. As before we denote this order by p_0 and call it the *reduced order*. For t-dependent coefficients we are no longer able to compute p_0. We therefore *assume* that the function $Q(s, t_0, t)$ admits the estimate

$$| Q(s, t_0, t) | \leqslant C_1(1 + | s |^{mp}) \exp[C(t - t_0) | s |^{p_0}]$$

(this is obviously true for $p_0 = p$). Then the approach which was described for the case of constant coefficients can be carried through without modifications and leads to the same result. In particular, for $p_0 > 1$ *the uniqueness class of the Cauchy problem*

$$\frac{\partial u(x, t)}{\partial t} = P\left(i \frac{\partial}{\partial x}, t\right) u(x, t), \qquad u(x, 0) = u_0(x), \tag{4}$$

is characterized by the inequality

$$| f(x) | \leqslant C_1 \exp[b_0 | x |^{p_0'}],$$

where p_0 is the reduced order of the system (4) and $(1/p_0) + (1/p_0') = 1$.

5. Examples

5.1. The Equation $u_t = au_{xx}$

We again consider the equation

$$\frac{\partial u}{\partial t} = a \frac{\partial^2 u}{\partial x^2}, \tag{1}$$

where a is a complex constant. In particular, for $a > 0$ one obtains the heat equation, for $a < 0$, one obtains the "inverse heat equation," and for purely imaginary a one obtains the Schrödinger equation, describing the one-dimensional motion of a free particle in quantum mechanics.

We shall determine the uniqueness class for the Cauchy problem for Eq. (1). Replacing $i(\partial/\partial x)$ by s we have the equation

$$dv/dt = -as^2 v$$

with the resolvent function $Q(s, t_0, t) = \exp(-as^2(t - t_0))$. This function is of order 2.

The number p_0' is in this case also equal to 2. According to Theorem 1 in Section 3, there can exist only one solution of Eq. (1) within the class of functions $f(x)$ which satisfy the inequality

$$|f(x)| \leqslant C \exp[bx^2].$$

In the present case one cannot improve the results of Theorem 1 in Section 3 by replacing the exponent 2 by $2 + \epsilon$, for any positive ϵ. Indeed, Eq. (1) admits the solution

$$u(x, t) = \sum_{m=0}^{\infty} a^m \frac{f^{(m)}(t)}{(2m)} x^{2m} \tag{2}$$

if the function $f(t)$ satisfies the inequalities

$$|f^{(m)}(t)| \leqslant \epsilon_m{}^m (2m)!, \qquad \epsilon_m \to 0, \tag{3}$$

which guarantee the convergence of the series (2) for all values of x. Setting $\epsilon_m = m^{\delta-1}$, with $1 > \delta > 0$, it follows that $\epsilon_m{}^m(2m)!$ is of order $m^{m(1+\delta)}$, and on the basis of the Carleman-Ostrowski theorem,[9] there exists a function $f(x) \not\equiv 0$, satisfying the inequalities (3) and vanishing together with all its derivatives for $t = 0$. The corresponding solution $u(x, t)$ vanishes identically for $t = 0$, and will differ from zero for $t > 0$. In order to estimate the growth of this function as $|x| \to \infty$, we substitute the inequality (3) into Eq. (2), obtaining

$$|u(x, t)| \leqslant \sum_{m=0}^{\infty} a^m \epsilon_m{}^m |x|^{2m} = \sum_{m=0}^{\infty} a^m m^{m(\delta-1)} |x|^{2m} \leqslant C \exp[b|x|^{2/(1-\delta)}]$$

(cf. a similar estimation in Section 2, Chapter IV, Volume 2). For sufficiently small δ the exponent $2/(1 - \delta)$ becomes smaller than $2 + \epsilon$, for any $\epsilon > 0$. Thus, for any $\epsilon > 0$ the Cauchy problem for Eq. (1) is no longer uniquely solvable in the class of functions satisfying the inequality

$$|f(x)| \leqslant C \exp[b |x|^{2+\epsilon}].$$

[9] Cf. Section 7, Chapter IV, Volume 2.

5.2. The Equation $u_{tt} = a^2 u_{xx}$

We consider the equation

$$\frac{\partial^2 u}{\partial t^2} = a^2 \frac{\partial^2 u}{\partial x^2}, \tag{1}$$

where a is a (complex) constant. For real a, in particular, the wave equation is obtained, and for imaginary a one obtains the Laplace equation. We determine the uniqueness class for the Cauchy problem

$$u(x, 0) = u_0(x), \qquad \frac{\partial u(x, 0)}{\partial t} = \bar{u}_0(x).$$

Defining the new unknown functions $u_1 = u$, $u_2 = \partial u/\partial t$ we replace Eq. (1) by the system

$$\frac{\partial u_1}{\partial t} = u_2, \qquad \frac{\partial u_2}{\partial t} = a^2 \frac{\partial^2 u_1}{\partial x^2}. \tag{2}$$

Replacing further $i(\partial/\partial x)$ by s we obtain the system

$$\frac{dv_1}{dt} = v_2, \qquad \frac{dv_2}{dt} = -a^2 s^2 v.$$

The fundamental matrix of solutions has the form

$$Q(s, t_0, t) = \begin{Vmatrix} \cos as(t - t_0) & \frac{1}{as} \sin as(t - t_0) \\ -as \sin as(t - t_0) & \cos as(t - t_0) \end{Vmatrix}.$$

The elements of this matrix are entire functions of s, of order $p_0 = 1$ (here $p_0 = 1, p = 2$). Thus according to Theorem 1, Section 3, we can assert that the uniqueness of the solution of the Cauchy problem for Eq. (1) is guaranteed within the class of functions $f(x)$ satisfying the inequality

$$|f(x)| \leqslant C \exp[|x|^{\lambda}],$$

with arbitrary λ.

If a is a real constant the result can be improved. In this case, for real $s = \sigma$ the matrix $Q(s, t_0, t)$ increases not faster than a polynomial (of first degree) in s. Therefore, according to the definition of Section 4.3, the system (2) is hyperbolic, and the uniqueness of its solution is guaranteed in the class of all functions, without any restrictions on their growth.

5.3. The Equation $u_{tt} = (1/a)\, u_x$

We consider the equation

$$\frac{\partial^2 u}{\partial t^2} = \frac{1}{a}\frac{\partial u}{\partial x} \tag{1}$$

with the initial conditions

$$u(x, 0) = u_0(x), \qquad \frac{\partial u(x, 0)}{\partial t} = \bar{u}_0(x).$$

This equation is of the same type as the one in Example 5.1, but the Cauchy problem has now a completely different meaning. Setting $a = 1$ and letting t be a spatial coordinate and x a time coordinate, the corresponding heat equation will determine the temperature $u(x, t)$ if its values are known at all times at one point and the values of its spatial derivative are known for all times at the same point.

In order to find the uniqueness class for the Cauchy problem for Equation (1) we replace it by a system, by means of the substitutions $u_1 = u$, $u_2 = \partial u / \partial t$:

$$\frac{\partial u_1}{\partial t} = u_2\,, \qquad \frac{\partial u_2}{\partial t} = \frac{1}{a}\frac{\partial u_1}{\partial x}\,.$$

Setting $s = i(\partial / \partial x)$, we obtain

$$\frac{dv_1}{dt} = v_2\,, \qquad \frac{dv_2}{dt} = \frac{s}{ai} v_1\,.$$

The resolvent matrix is of the form

$$Q(s, t_0, t) = \begin{Vmatrix} \cos(is/a)^{1/2}(t - t_0) & (a/is)^{1/2} \sin (is/a)^{1/2} (t - t_0) \\ -(is/a)^{1/2} \sin(is/a)^{1/2}(t - t_0) & \cos(is/a)^{1/2}(t - t_0) \end{Vmatrix}.$$

The matrix elements here are entire functions of order $p_0 = \frac{1}{2}$. It then follows from Theorem 1, Section 3 that *the uniqueness of the solution of the Cauchy problem for equation* (1) *is guaranteed within the class of all functions without any growth restrictions.*

In the following section we show how to simplify the computations involved in establishing the uniqueness class by making use of the characteristic roots (zeros) of the corresponding system.

6. The Connection between the Reduced Order of a System and Its Characteristic Roots

6.1. A Fundamental Inequality

As we have seen, the reduced order plays an essential role in the construction of the uniqueness class of solutions of the Cauchy problem for the system

$$\frac{\partial u(x, t)}{\partial t} = P\left(i \frac{\partial}{\partial x}\right) u(x, t). \tag{1}$$

By definition, this reduced order is the order of the entire matrix-function

$$Q(s, t) = e^{tP(s)}.$$

in the complex s-plane.

In this section we shall reduce the problem of finding the order of exponential growth of the matrix $Q(s, t)$ to the simpler problem of finding the power-law order of increase of the characteristic roots of the matrix $P(s)$. We denote by $\lambda_1(s),\ldots, \lambda_m(s)$ the roots of the characteristic equation

$$\det \| P(s) - \lambda E \| = 0. \tag{2}$$

The functions $\lambda_1(s),\ldots, \lambda_m(s)$ are called the *characteristic roots* of the system (1). They are defined and continuous for all complex values of s.

The characteristic roots of the system (1) occur in the solution of the following problem: *find solutions of the system*

$$\frac{\partial u_j(x, t)}{\partial t} = \sum_{k=1}^{m} P_{jk}\left(i \frac{\partial}{\partial x}\right) u_k(x, t) \qquad (j = 1, 2,\ldots, m), \tag{3}$$

which are of the form

$$u_j(x, t) = C_j \exp[\lambda t - is_1x_1 - \cdots - is_nx_n] \qquad (j = 1, 2,\ldots, m), \tag{4}$$

with fixed values of $s_1 ,\ldots, s_n$ and λ.

Indeed, substituting (4) into (3) and dividing by $\exp[\lambda t - is_1x_1 - \cdots - is_nx_n]$ we have

$$\lambda C_j = \sum_{k=1}^{m} C_kP_{jk}(s) \qquad (j = 1, 2,\ldots, m).$$

This system admits nonvanishing solutions only if

$$\det \| P(s) - \lambda E \| = 0,$$

i.e., if λ equals one of the characteristic roots of the matrix $P(s)$.

We define

$$\Lambda(s) = \max_k \operatorname{Re} \lambda_k(s). \tag{5}$$

The main result of this section is summarized in the following

Theorem 1. *For any $m \times m$ matrix $P(s)$, the elements of which are polynomials of $s = (s_1, \ldots, s_n)$ of degrees $\leqslant p$, the following estimate holds for $t \geqslant 0$:*

$$e^{t\Lambda(s)} \leqslant \| e^{tP(s)} \| \leqslant C\,(1 + |s|)^{(m-1)p} e^{t\Lambda(s)}, \tag{6}$$

where $\| e^{tP(s)} \|$ denotes the norm of the matrix $e^{tP(s)}$ (i.e., the norm of the corresponding linear operator in m-dimensional Euclidean space).

The proof of this theorem is based on the following lemma.

Lemma 1. *Let P be an arbitrary $m \times m$ matrix with complex elements. Let further $\lambda_1, \ldots, \lambda_m$ denote the characteristic roots (eigenvalues) of the matrix P and let $\Lambda = \max \operatorname{Re} \lambda_j$. Then for $t \geqslant 0$ we have the inequality*

$$\| e^{tP} \| \leqslant e^{t\Lambda}(1 + 2t \| P \| + \cdots + (2t)^{m-1} \| P \|^{m-1}). \tag{7}$$

Proof. If $f(\lambda) = \sum_{k=0}^{\infty} a_k \lambda^k$ is an entire analytic function of λ and P a matrix, then $f(P)$ is defined by the power series

$$f(P) = \sum_{k=0}^{\infty} a_k P^k.$$

But the matrix $f(P)$ can also be constructed as a polynomial $R(P)$. In order to achieve this[10] one has to take a polynomial $R(\lambda)$, which takes at the points $\lambda_1, \ldots, \lambda_m$ corresponding to the characteristic roots of the matrix P the values $f_1 = f(\lambda_1), \ldots, f_m = f(\lambda_m)$. Here we assume that the zeros are all different (nondegenerate eigenvalues). The case of multiple zeros will be considered later.

In order to determine the polynomial $R(\lambda)$ we choose from among the many available interpolation formulas the Newton interpolation polynomial

$$R(\lambda) = b_1 + b_2(\lambda - \lambda_1) + b_3(\lambda - \lambda_1)(\lambda - \lambda_2) + \cdots + b_m(\lambda - \lambda_1) \ldots (\lambda - \lambda_{m-1}). \tag{8}$$

[10] Cf. p. 83 of the book by Gantmakher, cited on p. 58.

The coefficients $b_1, \ldots, b_m$ are determined by putting successively $\lambda = \lambda_1, \lambda_2, \ldots, \lambda_m$. We thus obtain the system of equations

$$\begin{aligned} f_1 &= b_1, \\ f_2 &= b_1 + b_2(\lambda_2 - \lambda_1), \\ f_3 &= b_1 + b_2(\lambda_3 - \lambda_1) + b_3(\lambda_3 - \lambda_1)(\lambda_3 - \lambda_2), \\ &\cdots \\ f_m &= b_1 + b_2(\lambda_m - \lambda_1) + b_3(\lambda_m - \lambda_1)(\lambda_m - \lambda_2) + \cdots \\ &\quad + b_m(\lambda_m - \lambda_1) \ldots (\lambda_m - \lambda_{m-1}). \end{aligned} \tag{9}$$

We introduce the notations

$$[f_j] = f_j, \qquad [f_{j_1 j_2}] = \frac{[f_{j_2}] - [f_{j_1}]}{\lambda_{j_2} - \lambda_{j_1}},$$

$$[f_{j_1 j_2 j_3}] = \frac{[f_{j_1 j_3}] - [f_{j_1 j_2}]}{\lambda_{j_3} - \lambda_{j_2}},$$

$$\cdots\cdots\cdots$$

$$[f_{j_1 j_2 \cdots j_k}] = \frac{[f_{j_1 \cdots j_{k-2} j_k}] - [f_{j_1 \cdots j_{k-2} j_{k-1}}]}{\lambda_{j_k} - \lambda_{j_{k-1}}}.$$

Then the Eqs. (9) yield successively

$$\begin{aligned} b_1 &= f_1 = [f_1], \\ b_2 &= \frac{f_2 - f_1}{\lambda_2 - \lambda_1} = [f_{12}], \\ b_3 &= \frac{f_3 - f_1 - [f_{12}](\lambda_3 - \lambda_1)}{(\lambda_3 - \lambda_1)(\lambda_3 - \lambda_2)} = \frac{[f_{13}] - [f_{12}]}{\lambda_3 - \lambda_2} = [f_{123}], \\ &\cdots\cdots\cdots \\ b_{k+1} &= \{f_{k+1} - f_1 - [f_{12}](\lambda_{k+1} - \lambda_1) - \cdots \\ &\quad - [f_{12\ldots k}](\lambda_{k+1} - \lambda_1) \cdots (\lambda_{k+1} - \lambda_k)\} \\ &\quad : \{(\lambda_{k+1} - \lambda_1)(\lambda_{k+1} - \lambda_2) \cdots (\lambda_{k+1} - \lambda_k)\} \\ &= \{[f_{1,k+1}] - [f_{12}] - [f_{123}](\lambda_{k+1} - \lambda_2) - \cdots \\ &\quad - [f_{12\ldots k}](\lambda_{k+1} - \lambda_2) \cdots (\lambda_{k+1} - \lambda_k)\} \\ &\quad : \{(\lambda_{k+1} - \lambda_2) \cdots (\lambda_{k+1} - \lambda_k)\} \\ &= \{[f_{12,k+1}] - [f_{123}] - \cdots \\ &\quad - [f_{12\ldots k}](\lambda_{k+1} - \lambda_3) \cdots (\lambda_{k+1} - \lambda_k)\} \\ &\quad : \{(\lambda_{k+1} - \lambda_3) \cdots (\lambda_{k+1} - \lambda_k)\} = [f_{12 \cdots k, k+1}], \\ &\cdots\cdots\cdots \\ b_m &= [f_{12 \cdots m}]. \end{aligned}$$

One can obtain estimates for these quantities by means of complex integrations. We define the expressions ($k = 1, 2,\dots, m$)

$$u_k(\lambda) = \int_0^1 \int_0^{t_1} \cdots \int_0^{t_{k-1}} f^{(k)}[\lambda_1 + (\lambda_2 - \lambda_1)t_1 + \cdots + (\lambda_k - \lambda_{k-1})t_{k-1} + (\lambda - \lambda_k)t_k]\, dt_1 \cdots dt_k .$$

Integrating over the coordinate t_k we have

$$\begin{aligned} u_k(\lambda) &= \frac{1}{\lambda - \lambda_k} \int_0^1 \int_0^{t_1} \cdots \int_0^{t_{k-2}} f^{(k-1)}[\lambda_1 + (\lambda_2 - \lambda_1)t_1 + \cdots \\ &\quad + (\lambda_k - \lambda_{k-1})t_{k-1} + (\lambda - \lambda_k)t_k]\Big|_0^{t_{k-1}} dt_1 \cdots dt_{k-1} \\ &= \frac{1}{\lambda - \lambda_k} \left\{ \int_0^1 \int_0^{t_1} \cdots \int_0^{t_{k-2}} f^{(k-1)}[\lambda_1 + (\lambda_2 - \lambda_1)t_1 + \cdots \right. \\ &\quad + (\lambda - \lambda_{k-1})t_{k-1}]\, dt_1 \cdots dt_{k-1} \\ &\quad - \int_0^1 \int_0^{t_1} \cdots \int_0^{t_{k-2}} f^{(k-1)}[\lambda_1 + (\lambda_2 - \lambda_1)t_1 + \cdots \\ &\quad \left. + (\lambda_k - \lambda_{k-1})t_{k-1}]\, dt_1 \cdots dt_{k-1} \right\} = \frac{u_{k-1}(\lambda) - u_{k-1}(\lambda_k)}{\lambda - \lambda_k}. \end{aligned} \tag{10}$$

We further denote $u_0(\lambda) = f(\lambda)$. Then (10) yields successively

$$u_0(\lambda_1) = [f_1],$$

$$u_1(\lambda_2) = \frac{[f_2] - [f_1]}{\lambda_2 - \lambda_1} = [f_{12}],$$

$$\cdots$$

$$u_{m-1}(\lambda_m) = \frac{[f_{12\ldots m-2,m}] - [f_{12\ldots m-2,m-1}]}{\lambda_m - \lambda_{m-1}} = [f_{12\ldots m}].$$

Hence the numbers $u_{k-1}(\lambda_k)$ ($k = 1, 2,\dots, m$) coincide with the coefficients b_k of the required interpolating polynomial $R(\lambda)$, and thus

$$b_k = \int_0^1 \int_0^{t_1} \cdots \int_0^{t_{k-2}} f^{(k-1)}[\lambda_1 + (\lambda_2 - \lambda_1)t_1 + \cdots + (\lambda_k - \lambda_{k-1})t_{k-1}]\, dt_1 \cdots dt_{k-1} . \tag{11}$$

We now show that for all values of $t_1, t_2, \dots, t_{k-1}$ which satisfy the inequalities $0 \leqslant t_{k-1} \leqslant t_{k-2} \cdots t_1 \leqslant 1$, the argument of the function $f^{(k-1)}$ is situated within the smallest convex polygon B containing the points $\lambda_1, \lambda_2, \dots, \lambda_k$.

We have, in fact:

$$\lambda_1 + (\lambda_2 - \lambda_1)t_1 + \cdots + (\lambda_k - \lambda_{k-1})t_{k-1}$$
$$= \lambda_1(1 - t_1) + \lambda_2(t_1 - t_2) + \cdots + \lambda_k t_{k-1} ;$$

the coefficients in front of $\lambda_1, \lambda_2, \ldots, \lambda_k$ are nonnegative and their sum is

$$1 - t_1 + t_1 - t_2 + \cdots + t_{k-2} - t_{k-1} + t_{k-1} = 1.$$

Thus the argument of the function $f^{(k-1)}$ coincides with the center of mass of the system of masses $1 - t_1, t_1 - t_2, \ldots, t_{k-1}$ situated respectively in the points $\lambda_1, \lambda_2, \ldots, \lambda_k$ and consequently is inside the polygon B.

Let $M_k = \max_B |f^{(k)}(\lambda)|$. Then from (11) we obtain the final estimate

$$| b_k | \leqslant M_{k-1} . \tag{12}$$

In particular, for a function $f(\lambda) = e^{t\lambda}$, we have for $t \geqslant 0$

$$M_k = \max_B | (e^{t\lambda})^{(k)} | = t^k \max_B | e^{t\lambda} | = t^k e^{t\Lambda},$$

with Λ the maximal real part of the characteristic roots $\lambda_1, \lambda_2, \ldots, \lambda_m$. It follows therefore for the matrix P

$$\begin{aligned} \| e^{tP} \| &= \| R(P) \| = \| b_1 + b_2(P - \lambda_1 E) \\ &\quad + b_3(P - \lambda_1 E)(P - \lambda_2 E) + \cdots \| \\ &\leqslant M_0 + M_1 \| P - \lambda_1 E \| + M_2 \| P - \lambda_1 E \| \| P - \lambda_2 E \| + \cdots \\ &\leqslant e^{t\Lambda}[1 + t(\| P \| + | \lambda_1 |) + t^2(\| P \| + | \lambda_1 |)(\| P \| + | \lambda_2 |) + \cdots] \\ &\leqslant e^{t\Lambda}[1 + 2t \| P \| + 4t^2 \| P \|^2 + \cdots + (2t \| P \|)^{m-1}], \end{aligned} \tag{13}$$

since any eigenvalue of a matrix cannot be larger in absolute value than the norm of the matrix. This estimate was obtained under the assumption that the characteristic roots of the matrix P were simple. Continuity considerations show that the estimate is also valid in the general case, when there are multiple zeros among the characteristic roots. This accomplishes the proof of the lemma.

Proof of Theorem 1. Let us apply the results of the lemma to the case $P = P(s)$. Since the square of the norm of a matrix is not larger that the sum of the absolute squares of the matrix elements, and the polynomials which make up the matrix $P(s)$ are of maximal degree p, we obtain

$$\| P \| = \| P(s) \| \leqslant C_1(1 + | s |)^p.$$

Consequently we arrive at the final formula

$$\| Q(s, t) \| \leqslant \| e^{tP(s)} \| \leqslant C(1 + | s |)^{p(m-1)} e^{t\Lambda(s)}, \qquad t \geqslant 0,$$

which is the same as the right-hand side of the inequality (6).

In order to obtain the left-hand side of the inequality (6), we consider a normalized eigenvector u_j of the matrix $Q(s, t)$ belonging to the eigenvalue $e^{t\lambda_j(s)}$ for which $\operatorname{Re} \lambda_j(s) = \Lambda(s)$. Then

$$Q(s, t)u_j = e^{t\lambda_j(s)} u_j ,$$

whence

$$\| Q(s, t) \| = \sup_{\| u \| = 1} \| Q(s, t)u \| \geqslant \| Q(s, t)u_j \| = e^{t\Lambda(s)},$$

as required.

Theorem 1 simply reduces the estimation of the growth of the function $e^{tP(s)}$ to an estimate of the growth of $\Lambda(s)$, where $\Lambda(s)$ is the maximum of the real parts of the characteristic zeros of the matrix $P(s)$.

We shall call (*exact*) *degree of growth* (or power of growth) of a function $f(s)$ the infimum of the numbers ρ which have the property that the following inequality is verified for all s:

$$| f(s) | \leqslant C_\rho (1 + | s |)^\rho. \tag{14}$$

The inequality (6) immediately implies that the (*exact*) *degree of growth of the function* $\Lambda(s)$ *is equal to the* (*exact*) *exponential order of growth of the matrix* $e^{tP(s)}$ *and thus coincides with the reduced order of the system* (1).

It is clear from the second definition of the characteristic roots (given at the beginning of this section) that these roots are unchanged if the functions $u_j(x, t)$ are replaced by their derivatives and their linear combinations. This implies the invariance of the reduced order of the system with respect to such transformations.

Note. The inequality (6) allows one to reformulate in terms of the function $\Lambda(s)$ some of the properties of the entire function $e^{tP(s)}$, properties which are consequences of the classical theory of the growth of entire functions, such as the Phragmén-Lindelöf principle, the indicatrix theory, etc.

Let us illustrate one of these properties. It is well known that an entire function $f(s)$ of order p_0, which decreases exponentially in some direction with the majorization

$$| f(s) | \leqslant C \exp[-| s |^{p_1},] \qquad p_1 > p_0 ,$$

vanishes identically.

It follows then, from our previous results, that there does not exist a matrix $P(s) = \| P_{jk}(s)\|$, for which the function $\Lambda(s)$ satisfies in some direction an inequality

$$\Lambda(s) \leqslant -|s|^{p_1}, \qquad p_1 > p_0 .$$

If the contrary were true, the function $e^{tP(s)}$ would vanish identically on the basis of the foregoing, which is impossible.

6.2. Computation of the Number p_0

The question how to determine the degree of growth of the function $\Lambda(s)$ naturally arises directly in terms of the matrix $P(s)$. The following theorem answers the question:

Theorem 2. *Consider the expansion of the characteristic polynomial of the matrix $P(s)$ in powers of λ:*

$$\det \| P(s) - \lambda E \| = (-\lambda)^m + P_1(s)\lambda^{m-1} + \cdots + P_m(s).$$

Let p_j denote the degree of the polynomial $P_j(s)$ (with respect to the totality of variables $s_1, \ldots, s_n$). Then the degree of growth p_0 of the function $\Lambda(s)$ can be computed by means of the formula

$$p_0 = \max_{1 \leqslant j \leqslant m} \frac{p_j}{j} .$$

The following functions

$$\bar{\Lambda}(s) = \max_j | \operatorname{Re} \lambda_j(s) |,$$

$$\Pi(s) = \max_j | \operatorname{Im} \lambda_j(s) |,$$

$$M(s) = \max_j | \lambda_j(s) |.$$

have the same degree of growth.

We shall carry out the *proof* in several steps.

I. Let us set temporarily

$$\max_j \frac{p_j}{j} = q_0 .$$

We show that the roots $\lambda_j(s)$ of the characteristic polynomial verify the inequality

$$| \lambda_j(s) | \leqslant C(1 + \rho)^{q_0} \qquad (\rho = | s |). \tag{1}$$

Admitting that the contrary is true, we find a sequence of points $\{s_\nu\}$ $|s_\nu| \to \infty$ as $\nu \to \infty$, for which one of the roots, say the first for the sake of definiteness, increases in such a manner, that

$$|\lambda_1(s_\nu)| > \nu(1+\rho_\nu)^{q_0} \qquad (\rho_\nu = |s_\nu|)$$

holds. We have

$$\begin{aligned} 0 &= \lambda_1^m(s_\nu)(-1)^m + P_1(s_\nu)\lambda_1^{m-1}(s_\nu) + \cdots + P_m(s_\nu) \\ &= \lambda_1^m(s_\nu)\left[(-1)^m + \frac{P_1(s_\nu)}{\lambda_1(s_\nu)} + \cdots + \frac{P_m(s_\nu)}{\lambda_1^m(s_\nu)}\right]. \end{aligned}$$

We obtain an immediate contradiction if all the ratios inside the square bracket vanish as $\nu \to \infty$. Consequently there exists a j, and a subsequence $\{s_{\nu_k}\}$, which we denote again by $\{s_\nu\}$, such that

$$|P_j(s_\nu)|/|\lambda_1^j(s_\nu)| > C > 0.$$

Hence

$$|P_j(s_\nu)| > C|\lambda_1^j(s_\nu)| > C\nu^j(1+\rho_\nu)^{q_0 j}$$

and consequently the degree p_j of the polynomial $P_j(s)$ is higher than $q_0 j$:

$$p_j > q_0 j.$$

But then

$$p_j/j > q_0,$$

and thus even more strongly

$$\max(p_j/j) > q_0,$$

which contradicts the definition of q_0. Thus the inequality (1) is established. At the same time we have established the inequalities

$$\operatorname{Re}\lambda_j(s) \leqslant C(1+\rho)^{q_0}, \qquad |\operatorname{Re}\lambda_j(s)| \leqslant C(1+\rho)^{q_0},$$
$$|\operatorname{Im}\lambda(s)| \leqslant C(1+\rho)^{q_0},$$

so that the degrees of growth of the functions $\Lambda(s)$, $\bar{\Lambda}(s)$, $\Pi(s)$, and $M(s)$ are not larger than q_0.

II. The problem consists further in proving that the functions $\Lambda(s)$, $\bar{\Lambda}(s)$, $\Pi(s)$, and $M(s)$ have the same degree of growth. Before atacking this problem we establish the following lemma.

Lemma 2. *The real part* $Q_r(\sigma_1, \ldots, \tau_n)$ *and the imaginary part* $Q_i(\sigma_1, \ldots, \tau_n)$ *of the polynomial* $Q(s_1, \ldots, s_n)$ *of degree* k *are polynomials of the same degree* k *in the totality of variables* $\sigma_1, \ldots, \tau_n$; *for sufficiently large* $\rho \geqslant \rho_0$ *they satisfy the inequalities*

$$\begin{aligned} C_1\rho^k &\leqslant \max_{|s| \leqslant \rho} Q_r(s) \leqslant C_2\rho^k, \\ C_3\rho^k &\leqslant \max_{|s| \leqslant \rho} |Q_r(s)| \leqslant C_4\rho^k, \\ C_5\rho^k &\leqslant \max_{|s| \leqslant \rho} |Q_i(s)| \leqslant C_6\rho^k. \end{aligned} \tag{2}$$

Proof. It is obvious that the degrees of the polynomials $Q_r(s)$ and $Q_i(s)$ in the ensemble of variables σ, τ cannot be higher than k, which makes the right halves of the inequalities (2) obviously true. We further separate in the polynomial Q the group of terms of highest degree k:

$$Q(s) = Q^0(s) + Q^1(s), \tag{3}$$

here $Q^1(s)$ is a polynomial of degree not higher than $k - 1$, whereas all terms in $Q^0(s)$ are exactly of degree k. The following decompositions hold together with (3):

$$Q_r(s) = Q_r^0(s) + Q_r^1(s), \qquad Q_i(s) = Q_i^0(s) + Q_i^1(s),$$

where $Q_r^0(s) = \operatorname{Re} Q^0(s)$ is a homogeneous polynomial of degree k in the variables $\sigma_1, \ldots, \tau_n$ and $Q_r^1(s) = \operatorname{Re} Q^1(s)$ is a polynomial in the same variables of degree not higher than $k - 1$: the symbols $Q_i^0(s)$ and $Q_i^1(s)$ have a similar meaning. Let α denote a complex number satisfying the equation $\alpha^k = i$. Since $Q^0(s)$ is a homogeneous polynomial of degree k, we have

$$Q^0(\alpha s) = \alpha^k Q^0(s) = iQ^0(s).$$

Thus

$$\max_{|s| \leqslant \rho} |Q_r^0(s)| = \max_{|s| \leqslant \rho} |Q_i^0(s)|.$$

But, obviously

$$\max |Q^0(s)| = C_0\rho^k \leqslant \max |Q_r^0(s)| + \max |Q_i^0(s)|,$$

hence

$$\max |Q_r^0(s)| = \max |Q_i^0(s)| \geqslant \tfrac{1}{2}C_0\rho^k.$$

The lower-order terms $Q_r^1(s)$ and $Q_i^1(s)$ are of degree not higher than $k - 1$ and consequently cannot affect the preceding estimates. Thus the inequalities

$$\max |Q_r^0(s)| \geqslant C_3\rho^k, \qquad \max |Q_i^0(s)| \geqslant C_5\rho^k$$

are certainly true for sufficiently large ρ. We further select the number α in such a manner than $\alpha^k = -1$. Then

$$Q_r^0(\alpha s) = -Q_r^0(s).$$

This shows that the ranges of the functions $Q_r^0(s)$, $-Q_r^0(s)$ coincide in the domain $|s| \leqslant \rho$. Therefore

$$\max_{|s| \leqslant \rho} Q_r^0(s) = \max_{|s| \leqslant \rho} |Q_r^0(s)| \geqslant \tfrac{1}{2} C\rho^k.$$

The lower-order terms which make up $Q_r^1(s)$ cannot decisively influence this estimate and we obtain for sufficiently large $\rho \geqslant \rho_0$:

$$\max_{|s| \leqslant \rho} Q_r^0(s) \geqslant C_1 \rho^k,$$

as required.

III. We now show that the ratio of the functions $\Lambda^*(\rho) = \max_{s \leqslant \rho} \Lambda(s)$ and $\bar{\Lambda}^*(\rho) = \max_{|s| \leqslant \rho} \bar{\Lambda}(s)$ is bounded from above and below by positive constants as $|\rho| \to \infty$. For this purpose we consider the polynomial

$$P_1(s) = \lambda_1(s) + \cdots + \lambda_m(s),$$

which appears as the coefficient of λ^{m-1} in the expansion of $\det \| P(s) - \lambda E \|$ in powers of λ. Let k be the degree of this polynomial in the variables $s_1, \ldots, s_n$. The function $\sum \operatorname{Re} \lambda_j(s)$ is the real part of the polynomial $P_1(s)$. For sufficiently large $\rho > \rho_0$, we have on the basis of the lemma

$$C_1 \rho^k \leqslant \max_{|s| \leqslant \rho} \sum \operatorname{Re} \lambda_j(s) \leqslant C_2 \rho^k \qquad (\rho = |s|), \tag{4}$$

$$C_2 \rho^k \leqslant \max_{|s| \leqslant \rho} \left| \sum \operatorname{Re} \lambda_j(s) \right| \leqslant C_4 \rho^k. \tag{5}$$

We order the λ_j in such a manner that $\operatorname{Re} \lambda_1(s) \leqslant \operatorname{Re} \lambda_2(s) \leqslant \cdots \leqslant \operatorname{Re} \lambda_m(s)$. The quantity $\Lambda^*(\rho)$ equals $\max_{|s| \leqslant \rho} \operatorname{Re} \lambda_m(s)$ and $\bar{\Lambda}^*(\rho)$ equals either $\max_{|s| \leqslant \rho} \operatorname{Re} \lambda_m(s)$ or $\max_{|s| \leqslant \rho} \operatorname{Re} |\lambda_1(s)|$. In the first case $\Lambda^*(\rho) = \bar{\Lambda}^*(\rho)$. Let us consider now the second case. For $|s| \leqslant \rho$ we have

$$\sum_{j=1}^{m} \operatorname{Re} \lambda_j(s) \leqslant m \operatorname{Re} \lambda_m(s) \leqslant m\Lambda^*(\rho). \tag{6}$$

If we take in the left-hand side of the inequality the maximum with respect to $|s| \leqslant \rho$, then, on the basis of (4), we obtain the inequality

$$C_1 \rho^k \leqslant m\Lambda^*(\rho). \tag{7}$$

Let $\nu + 1$ denote (for given s) the first subscript for which $\operatorname{Re} \lambda_j(s)$ becomes positive. The inequality (5) then implies that for $|s| \leqslant \rho$

$$\operatorname{Re} \lambda_{\nu+1}(s) + \cdots + \operatorname{Re} \lambda_m(s) \geqslant | \operatorname{Re} \lambda_1(s) + \cdots + \operatorname{Re} \lambda_\nu(s) | - C_4\rho^k,$$

and hence

$$m \operatorname{Re} \lambda_m(s) \geqslant | \operatorname{Re} \lambda_1(s) | - C_4\rho^k. \tag{8}$$

From (6) and (8) we obtain for $|s| \leqslant \rho$

$$m\Lambda^*(\rho) \geqslant | \operatorname{Re} \lambda_1(s) | - C_4\rho^k.$$

Taking the maximum with respect to $|s| \leqslant \rho$ in the right-hand side, we obtain

$$m\Lambda^*(\rho) \geqslant \bar{\Lambda}^*(\rho) - C_4\rho^k.$$

With (7) this yields the inequality

$$\frac{\bar{\Lambda}^*(\rho)}{\Lambda^*(\rho)} \leqslant m + \frac{C_4\rho}{\Lambda^*(\rho)} \leqslant m + \frac{C_4}{C_1} m.$$

Thus the ratio

$$\bar{\Lambda}^*(\rho)/\Lambda^*(\rho)$$

remains bounded from above as $|\rho| \to \infty$. Since $\Lambda^*(\rho) \leqslant \bar{\Lambda}^*(\rho)$ it is also bounded from below, which completes the proof of our assertion.

IV. It will be shown now that the ratio of the functions

$$\Pi^*(\rho) = \max_{|s| \leqslant \rho} \Pi(s) \quad \text{and} \quad \bar{\Lambda}^*(\rho) = \max_{|s| \leqslant \rho} \bar{\Lambda}(s)$$

is bounded from above and below by positive constants for $\rho \to \infty$. Let us assume the opposite to the true, i.e., that the ratio

$$\Pi^*(\rho)/\bar{\Lambda}^*(\rho)$$

is unbounded for $\rho \to \infty$. If the functions $\mu_j(s) = |\operatorname{Im} \lambda_j(s)|$ are labeled in such a manner that $\mu_1(s) \geqslant \mu_2(s) \geqslant \cdots \geqslant \mu_m(s)$, then also the ratio

$$\max_{|s| \leqslant \rho} \mu_1(s)/\bar{\Lambda}^*(\rho) = \mu_1(s_\rho)/\bar{\Lambda}^*(\rho) = C_\rho\,.$$

will be unbounded. Let q be the largest of the numbers for which $(|s| \leqslant \rho)$

$$\frac{\mu_1(s) \cdots \mu_q(s)}{C_\rho{}^q[\bar{\Lambda}^*(\rho)]^q}$$

does not vanish at infinity. We show that for $q < m$

$$\frac{\mu_1(s) \cdots \mu_{q-1}(s)\mu_{q+1}(s)}{C_\rho{}^q[\bar{\Lambda}^*(\rho)]^q} \to 0 \qquad (|\,s\,| \leqslant \rho,\ \rho \to \infty).$$

If the contrary is true, then for a sequence $\{s_\nu\}\ |\,s_\nu\,| \leqslant \rho_\nu \to \infty$ we would have the inequality

$$\mu_1(s_\nu) \cdots \mu_{q-1}(s_\nu)\mu_{q+1}(s_\nu) \geqslant \alpha C^q_{\rho_\nu}[\bar{\Lambda}^*(\rho_\nu)]^q, \qquad \alpha > 0. \tag{9}$$

Since for any $|\,s\,| \leqslant \rho_\nu$ we have

$$\mu_j(s) \leqslant \mu_1(s) \leqslant \mu_1(s_{\rho_\nu}) = C_{\rho_\nu}\bar{\Lambda}^*(\rho_\nu),$$

so that (9) implies the inequality $\mu_{q+1}(s_\nu) \geqslant C_{q_\nu}\bar{\Lambda}^*(\rho_\nu)$. It would follow then, since $\mu_q(s_\nu) \geqslant \mu_{q+1}(s_\nu)$, that

$$\mu_1(s_\nu) \cdots \mu_q(s_\nu)\mu_{q+1}(s_\nu) \leqslant \alpha^{q+1}C^{q+1}_{\rho_\nu}[\bar{\Lambda}^*(\rho_\nu)]^{q+1}$$

in contradiction with the definition of q.

We consider now the coefficient $P_q(s)$ in the polynomial $\det\|P(s) - \lambda E\,\|$. Let $P_{qr}(s)$ and $P_{qi}(s)$ denote its real and imaginary parts, respectively. The coefficient $P_q(s)$ is a sum of products of q of the roots $\lambda_1(s),\ldots, \lambda_m(s)$ each. The product $\mu_1(s) \cdots \mu_q(s)$ is the absolute value of one of the terms of $P_{qr}(s)$ or of $P_{qi}(s)$. According to its construction, this product is larger than $\beta \cdot C_\rho{}^q[\bar{\Lambda}^*(\rho)]^q$, $\beta > 0$, for sufficiently large s, $|\,s\,| \leqslant \rho$. Any other term in the polynomials $P_{qr}(s)$ or $P_{qi}(s)$ is obtained from this product by replacing some factors with factors $\mu_j(s)$ $(j > q)$ or with $\operatorname{Re}\lambda_j(s)$. Consequently these terms increase considerably more slowly than the product under consideration. Therefore, one of the polynomials $P_{qr}(s)$ or $P_{qi}(s)$ increases faster than the other one. But this is in contradiction with the results of the lemma in Section 6.2. We thus reach the conclusion that the ratio $\Pi^*(\rho)/\bar{\Lambda}^*(\rho)$ is bounded from above. Replacing the polynomial

$$P(s, \lambda) = \lambda^m + P_1(s)\lambda^{m-1} + P_2(s)\lambda^{m-2} + P_3(s)\lambda^{m-3} + \cdots$$

with the polynomial

$$P'(s, \lambda) = \lambda^m + iP_i(s)\lambda^{m-1} - P_2(s)\lambda^{m-2} - iP_3(s)\lambda^{m-3} + \cdots,$$

for which the roots $\lambda_j'(s)$ are related to the roots of the original polynomial through the relation

$$\lambda_j'(s) = i\lambda_j(s),$$

we interchange the roles of the functions $\bar{\Lambda}(s)$ and $\Pi(s)$. This implies that the ratio $\bar{\Lambda}^*(\rho)/\Pi^*(\rho)$ is bounded from above, as stated.

V. Since the function $M^*(\rho) = \max_{|s|\leqslant\rho} M(s)$ satisfies the inequalities

$$\max\{\Pi^*(\rho), \bar{\Lambda}^*(\rho)\} \leqslant M^*(\rho) \leqslant \Pi^*(\rho) + \bar{\Lambda}^*(\rho),$$

it is clear that the ratios $M^*(\rho)/\Pi^*(\rho)$ and $M^*(\rho)/\bar{\Lambda}^*(\rho)$ are bounded from above and below for $\rho \to \infty$. On the basis of III, this is also true for $M^*(\rho)/\Lambda^*(\rho)$. Consequently, the four functions $\Lambda^*(\rho)$, $\bar{\Lambda}^*(\rho)$, $\Pi^*(\rho)$, $M^*(\rho)$ have the same degree of (power-law) growth, which according to I cannot be higher than $q_0 = \max(p_j/j)$. It remains only to be shown that this degree is exactly equal to q_0 and for this it is sufficient to consider one of the above functions, say $M^*(\rho)$.

Let us assume that the roots $\lambda_k(s)$ grow for $|s| \to \infty$ slower than $|s|^{q_0}$, so that

$$|\lambda_k(s)| = \epsilon |s|^{q_0}, \qquad \epsilon \to 0.$$

According to Vieta's theorem, the coefficient $P_j(s)$ of λ^j is equal to the sum of products of j characteristic roots $\lambda_1(s) \cdots \lambda_j(s)$. Thus

$$|P_j(s)| \leqslant \epsilon_j |s|^{jq_0}, \qquad \epsilon_j \to 0.$$

It follows that the degree p_j of the polynomial $P_j(s)$ is *smaller* than jq_0 for any j. Consequently

$$p_j/j < q_0 \qquad (j = 1, 2, \ldots, m),$$

or

$$\max(p_j/j) < q_0$$

which contradicts the definition of q_0. Thus the degree of growth of the four functions $\Lambda(s)$, $\bar{\Lambda}(s)$, $\Pi(s)$, $M(s)$ equals q_0. In addition the following inequalities hold for $\Lambda(s)$

$$\Lambda(s) \leqslant C |s|^{q_0} \qquad \text{for all } |s| \geqslant \rho_0,$$

$$\Lambda(s) \geqslant C_1 |s|^{q_0} \qquad \text{for some sequence } \{s_\nu\} \ |s_\nu| \to \infty.$$

This accomplishes the proof of our theorem.

Corollary 1. *The reduced order of any system of type* (1), *Section* 6.1, *is a rational number with the denominator not larger than the number of equations in the system. For one equation*

$$\frac{\partial u}{\partial t} = P\left(i\frac{\partial}{\partial x}\right) u$$

the reduced order is an integer and is equal to the order l of the system.

Corollary 2. *If the characteristic roots increase slower than any positive power of $|s|$ these roots are constant.*

Indeed, in the case under consideration $q_0 = 0$, consequently $p_j = 0$ for any j. Thus the characteristic polynomial has coefficients which do not depend on s. But then the zeros of that polynomial are also independent of s.

6.3. Computation of the Reduced Order of Systems with Higher-Order Derivatives with Respect to t

We consider the system of partial differential equations containing partial derivatives of higher order with respect to t:

$$\frac{\partial^{\mu_j} u_j(x, t)}{\partial t^{\mu_j}} = \sum_{k=1}^{m} P_{jk}\left(\frac{\partial}{\partial t}, i\frac{\partial}{\partial x_1}, \dots, i\frac{\partial}{\partial x_n}\right) u_k(x, t) \quad (j = 1, 2, \dots, m), \tag{1}$$

where the degree of the polynomial P_{jk} in the variable $\partial/\partial t$ is smaller than the number μ_j. As established, such a system can be reduced to a system which is of first order in t, by introducing new unknown functions as follows:

$$\begin{aligned}
u_{11} &= u_1, & u_{12} &= \frac{\partial u_1}{\partial t}, \dots, u_{1\mu_1} = \frac{\partial^{\mu_1 - 1} u_1}{\partial t^{\mu_1 - 1}},\\
u_{21} &= u_2, & u_{22} &= \frac{\partial u_2}{\partial t}, \dots, u_{2\mu_2} = \frac{\partial^{\mu_2 - 1} u_2}{\partial t^{\mu_2 - 1}},\\
u_{m1} &= u_m, & u_{m2} &= \frac{\partial u_m}{\partial t}, \dots, u_{m\mu_m} = \frac{\partial^{\mu_m - 1} u}{\partial t^{\mu_m - 1}}.
\end{aligned}$$

This substitution reduces the system (1) to the form

$$\begin{aligned}
&\frac{\partial u_{11}}{\partial t} = u_{12}, \qquad \frac{\partial u_{12}}{\partial t} = u_{13},\\
&\quad \dots\\
&\frac{\partial u_{1\mu_1}}{\partial t} = \sum P_{1k}\left(\frac{\partial}{\partial t}, i\frac{\partial}{\partial x_1}, \dots, i\frac{\partial}{\partial x_n}\right) u_{k1}, \qquad \frac{\partial u_{21}}{\partial t} = u_{22}\\
&\quad \dots\\
&\frac{\partial u_{2\mu_2}}{\partial t} = \sum P_{2k}\left(\frac{\partial}{\partial t}, i\frac{\partial}{\partial x_1}, \dots, i\frac{\partial}{\partial x_n}\right) u_{k1},\\
&\quad \dots\\
&\frac{\partial u_{m1}}{\partial t} = u_{m2}\\
&\quad \dots\\
&\frac{\partial u_{m\mu_m}}{\partial t} = \sum P_{mk}\left(\frac{\partial}{\partial t}, i\frac{\partial}{\partial x_1}, \dots, i\frac{\partial}{\partial x_n}\right) u_{k1},
\end{aligned} \tag{2}$$

where the t-derivatives of the functions u_{k1} in the right-hand sides are to be replaced by the corresponding functions u_{kl}.

In order to determine the characteristic roots (zeros) $\lambda_1(s),..., \lambda_m(s)$ of the system (2), it is not even necessary to write down the system completely and to form the determinant of order $\mu_1 + \cdots + \mu_m$. It is possible to prove that these roots can be determined directly from the system (1), as the solutions of the equation

$$\begin{vmatrix} P_{11}(\lambda, s) - \lambda^{\mu_1} & P_{12}(\lambda, s) & \cdots & P_{1m}(\lambda, s) \\ P_{21}(\lambda, s) & P_{22}(\lambda, s) - \lambda^{\mu_2} & \cdots & P_{2m}(\lambda, s) \\ & \cdot \ \cdot \ \cdot & & \\ P_{m1}(\lambda, s) & P_{m2}(\lambda, s) & \cdots & P_{mm}(\lambda, s) - \lambda^{\mu_m} \end{vmatrix} = 0. \tag{3}$$

For the proof we remember that (cf. Section 6.1) the characteristic roots of the system (2) in the unknown functions u_{jl}, are defined as the values of λ for which the system admits a solution of the form

$$\begin{aligned} u_{11} &= C_{11} \exp[\lambda t - is_1x_1 - \cdots - is_nx_n], \\ u_{12} &= C_{12} \exp[\lambda t - is_1x_1 - \cdots - is_nx_n], \\ &\cdots\cdots\cdots\cdots\cdots\cdots \\ u_{m\mu_m} &= C_{m\mu_m} \exp[\lambda t - is_1x_1 - \cdots - is_nx_n]. \end{aligned}$$

The equations which determine λ now take the form

$$\begin{aligned} &\lambda C_{11} = C_{12}, \quad \lambda C_{12} = C_{13}, \quad \lambda C_{1\mu_1} = \sum P_{1k}(\lambda, s)C_{k1}, \\ &\lambda C_{21} = C_{22}, \quad \lambda C_{22} = C_{23}, \quad \lambda C_{2\mu_2} = \sum P_{2k}(\lambda, s)C_{k1}, \\ &\cdots\cdots\cdots\cdots\cdots\cdots\cdots\cdots \\ &\lambda C_{m1} = C_{m2}, \quad \lambda C_{m2} = C_{m3}, \quad \lambda C_{m\mu_m} = \sum P_{mk}(\lambda, s)C_{k1}. \end{aligned} \tag{4}$$

Multiplying among themselves the equations within each group we obtain

$$\begin{aligned} \lambda^{\mu_1}C_{11} &= \sum_{k=1}^{m} P_{1k}(\lambda, s)C_{k1}, \\ \lambda^{\mu_2}C_{21} &= \sum_{k=1}^{m} P_{2k}(\lambda, s)C_{k1}, \\ &\cdots\cdots\cdots\cdots \\ \lambda^{\mu_m}C_{m1} &= \sum_{k=1}^{m} P_{mk}(\lambda, s)C_{k1}, \end{aligned} \tag{5}$$

It is clear form here that those values of λ which correspond to nonvanishing solutions for the C_{jk} (consequently, to nonvanishing solutions C_{j1}) are solutions of the system (3).

In particular, a single equation

$$\frac{\partial^m u}{\partial t^m} = P\left(\frac{\partial}{\partial t}, i\frac{\partial}{\partial x_1}, \ldots, i\frac{\partial}{\partial x_n}\right) u, \tag{6}$$

with the degree of the polynomial P in the variable $\partial/\partial t$ smaller than m, is equivalent to the system

$$\begin{aligned} \frac{\partial u_1}{\partial t} &= u_2, \\ \frac{\partial u_2}{\partial t} &= u_3, \\ &\cdots \\ \frac{\partial u_m}{\partial t} &= P\left(\frac{\partial}{\partial t}, i\frac{\partial}{\partial x_1}, \ldots, i\frac{\partial}{\partial x_n}\right) u_1, \end{aligned} \tag{7}$$

where the t-derivatives of u_1 in the last equation should be replaced by the appropriate u_j. The characteristic roots of the system (7) are solutions of the equation of mth degree

$$\lambda^m = P(\lambda, s_1, \ldots, s_n).$$

Thus for the wave equation in one dimension (Section 5, Example 2)

$$\frac{\partial^2 u}{\partial t^2} = a^2 \frac{\partial^2 u}{\partial x^2} \tag{8}$$

the characteristic equation has the form

$$\lambda^2 = -a^2 s^2$$

with the roots $\lambda_{1,2} = \pm ias$. The function $\Lambda(s)$ increases linearly in s and for real a is bounded on the real axis. Thus the wave equation is hyperbolic in the sense of Section 4.3.

For the equation

$$\frac{\partial^2 u}{\partial t^2} = \frac{1}{a}\frac{\partial u}{\partial x} \tag{9}$$

the characteristic equation has the form

$$\lambda^2 = -\frac{i}{a}s, \qquad \lambda_{1,2}(s) = \pm\left(-\frac{is}{a}\right)^{1/2},$$

Here the function $\Lambda(s)$ has the degree of growth $p_0 = \frac{1}{2}$. Hence the Cauchy problem is unique in the class of all functions, without any growth restrictions (cf. Section 5, Example 3).

The characteristic equation for

$$\frac{\partial^k u}{\partial t^k} = \frac{\partial^m u}{\partial x^m}$$

is

$$\lambda^k(s) = (-is)^m.$$

Thus the function $\Lambda(s)$ has a power law behavior, with the degree of growth m/k. We have indicated above that the reduced order p_0 is always rational. We now see conversely *that each rational number can play the role of reduced order.*

7. A Theorem of the Phragmén-Lindelöf Type

7.1. Formulation of the Problem and Examples

Any solution of the Cauchy-Riemann equation

$$\frac{\partial u}{\partial t} = i\frac{\partial u}{\partial x}, \tag{1}$$

has the following well-known property: If this solution satisfies the inequalities

$$|u(x, t)| \leqslant C \exp[(|t| + |x|)^\gamma] \qquad (\gamma < 1) \tag{2}$$

and

$$|u(x, 0)| \leqslant C(1 + |x|^h), \tag{3}$$

then $u(x, t)$ is a polynomial in the complex variable $x + it$. Indeed: The first inequality implies that as a function of $x + it$, $u(x, t)$ is of order $\leqslant \gamma < 1$, and the second inequality implies that along the real axis, and at the same time on the sides of an angle $\theta < \pi/\gamma$, this function increases not faster than a polynomial. Since $u(x, t)$ is an analytic function of the variable $x + it$, one of the corollaries to the Phragmén-Lindelöf theorem (cf. Section 7, Chapter IV, Volume 2) implies that $u(x, t)$ itself is a polynomial of degree $\leqslant h$.

The solutions of the equation

$$\frac{\partial u(x, t)}{\partial t} = \frac{\partial u(x, t)}{\partial x} \tag{4}$$

do not exhibit similar properties. Indeed, since the general solution of Eq. (4) is of the form $u(x, t) = U(x + t)$, where U is an arbitrary differentiable function, there exist special solutions which are bounded over the whole plane, but are polynomials neither in x nor in t.

The question naturally arises: for what systems of equations of the form:

$$\frac{\partial u_j(x, t)}{\partial t} = \sum_{k=1}^{m} P_{jk}\left(i\frac{\partial}{\partial x}\right) u_k(x, t) \qquad (j = 1, 2, \ldots, m) \tag{5}$$

do the solutions exhibit properties similar to those indicated for the solutions of Eq. (1)? The following theorem encompasses a wide class of such systems.

Theorem. *Let the characteristic roots of the matrix $\| P_{jk}(\sigma) \|$ be real for all real $\sigma = (\sigma_1, \ldots, \sigma_n)$ and let this matrix commute with its derivatives. If u_j ($j = 1, \ldots, m$) is a solution of the system* (5) *satisfying the following conditions:*

(*a*) *for all t and x:*[11]

$$| u_j(x, t) | \leqslant C \exp[a_0 \, | t |^\gamma + b_0 \, | x |^h], \qquad \gamma < 1, \qquad h < (2p_0)' \tag{6}$$

(*p_0 is the reduced of the system, cf. Section* 3.2)

(b)
$$| u_j(x, 0) | \leqslant C(1 + | x |^h), \tag{7}$$

then $u_j(x, t)$ are polynomials in t of degree $\leqslant r$:

$$u_j(x, t) = \sum_{k=0}^{r} U_{jk}(x) t^k, \qquad r \leqslant h + m + n + [1], \tag{8}$$

(*i.e., r is the smallest even number $\geqslant h + m + n$*) *and the functions $U_{jk}(x)$ are solutions of the system*

$$\begin{aligned} &\sum_{j=1}^{m} P_{jk}\left(i\frac{\partial}{\partial x}\right) U_{j,q-1}(x) = qU_{jq}(x) \qquad (q < r), \\ &\sum_{j=1}^{m} P_{jk}\left(i\frac{\partial}{\partial x}\right) U_{jr}(x) = 0. \end{aligned} \tag{9}$$

[11] In the n-dimensional case $\exp[b_0 \, | x |^h]$ is to be interpreted as

$$\exp[b_0 \, | x_1 |^h + \cdots + b_0 \, | x_m |^h].$$

The reality condition for the characteristic roots of the matrix $P(\sigma)$ takes on a particularly simple form in the case of a single equation ($m = 1$). In this case it simply means that the coefficients of this equation (involving the operators $i\,\partial/\partial x$) are real. Equation (1) satisfies this condition, but Eq. (4) does not.

The proof of this theorem is contained in the following section.

7.2. Proof of the Theorem

The condition (6) of the preceding section permits us to consider the solutions of the system (5) as generalized functions depending on a parameter t and defined over the test function space $W_q{}^q$ ($h < q < (2p_0)'$). By definition this test function space consists of entire analytic functions $\varphi(x) = \varphi(x_1, \ldots, x_n)$ which satisfy the inequality

$$| \varphi(x + iy) | \leqslant C \exp[- | ax |^q + | by |^q]$$

with constants a, b, C depending on φ (cf. Chapter I, Section 1).

Subjecting the system (5) of Section 7.1 to a Fourier transformation, we obtain the Fourier-dual system

$$\frac{dv_j(\sigma, t)}{dt} = \sum_{k=1}^{m} P_{jk}(\sigma) v_k(\sigma, t), \tag{1}$$

where $v_j(\sigma, t)$ is a functional on the Fourier-dual space $\widetilde{W}_q{}^q = W_{q'}^{q'}$ (Section 3, Chapter I). Any entire function of order $< q'$ is a multiplier in the space $W_{q'}^{q'}$.

In particular, the resolvent matrix of the system (5), Section 7.1

$$Q(s, t) = e^{tP(s)}$$

is of order p_0, as established in Section 3. Here p_0 is the reduced order of the system (5). Since $q < (2p_0)'$, it follows that $2p_0 < q'$, and even more so $p_0 < q'$, so that the resolvent matrix $Q(s, t)$ is a multiplier in the space $W_{q'}^{q'}$. As a consequence of the fundamental theorem in Section 4 the Fourier-transformed system

$$\frac{dv(s, t)}{dt} = P(s) v(s, t) \tag{2}$$

with the initial condition

$$v_0(\sigma) = \widetilde{u(x, 0)}$$

admits a unique solution in the space $[W_{q'}^{q'}]'$:

$$v(\sigma, t) = e^{tP(\sigma)}v_0(\sigma).$$

Since the generalized function $\widetilde{u(x, t)}$ is a solution of the system (2) and at the same time the Fourier transform of the solution $u(x, t)$ of the system (5) of Section 7.1, we have

$$v(\sigma, t) = \widetilde{u(x, t)} = e^{tP(\sigma)}v_0(\sigma).$$

By definition of the Fourier transform for any vector function $\varphi(x) \in W_q{}^q$ and $\psi(\sigma) = \widetilde{\varphi(x)} \in W_{q'}^{q'}$, we have

$$(e^{tP(\sigma)}v_0(\sigma), \psi(\sigma)) = (u(x, t), \varphi(x)). \tag{3}$$

We shall denote this function of t by $F_\varphi(t)$.

Since the initial vector-function $u(x, 0)$ satisfies the inequality (7) of Section 7.1, there exists an (even) number r such that the ratio

$$U(x) = u(x, 0)/(1 + x_1{}^2 + \cdots + x_n{}^2)^{r/2}$$

of the vector-function $u(x, 0)$ and the polynomial of degree r in the denominator, is an integrable function. Here r can be set equal to one of the two integers $\geqslant h + n + 1$ or $h + n + 2$, whichever is even. Then

$$u(x, 0) = (1 + x_1{}^2 + \cdots + x_n{}^2)^{r/2}U(x),$$

$$v_0(\sigma) = \widetilde{u(x, 0)} = \left(1 - \frac{\partial^2}{\partial\sigma_1{}^2} - \cdots - \frac{\partial^2}{\partial\sigma_n{}^2}\right)^{r/2} V(\sigma) = R\left(\frac{\partial}{\partial\sigma}\right) \widetilde{U(x)},$$

where $R(\partial/\partial\sigma)$ is a differential operator of order r. Making use of this, we transform the left-hand side of (3) according to the rules of handling vector-valued generalized functions:

$$\begin{aligned} F_\varphi(t) &= (e^{tP(\sigma)}v_0(\sigma), \psi(\sigma)) \\ &= (v_0(\sigma), e^{tP^*(\sigma)}\psi(\sigma)) \\ &= \left(R\left(\frac{\partial}{\partial\sigma}\right) V(\sigma), e^{tP^*(\sigma)}\psi(\sigma)\right) \\ &= \left(V(\sigma), R\left(-\frac{\partial}{\partial\sigma}\right) e^{tP^*(\sigma)}\psi(\sigma)\right) \\ &= \int_{-\infty}^{\infty} \overline{V(\sigma)} \left\{\sum_{k=0}^{r} t^k e^{tP^*(\sigma)}P_k(\sigma)R_k\left(\frac{\partial}{\partial\sigma}\right) \psi(\sigma)\right\} d\sigma. \end{aligned} \tag{4}$$

Since the matrix $P(\sigma)$ commutes with its derivatives, we have

$$\frac{\partial}{\partial \sigma_j} \exp[tP^*(\sigma)] = t \frac{\partial P^*(\sigma)}{\partial \sigma_j} \exp[tP^*(\sigma)]$$

etc., which proves the last equality in (4).

In order to prove that $F(t)$ is an entire analytic function of the complex variable t, of order smaller than 2, we show that the integral (4), as well as the integral of the derivative of the integrand with respect to t, converges absolutely for all complex t; we also give an estimate of the order.

Let $g_k(\sigma) = P_k(\sigma) R_k(\partial/\partial\sigma) \psi(\sigma)$. Both $\psi(\sigma)$ and $g_k(\sigma)$ belong to $W_{q'}^{q'}$. Hence we have the estimate

$$| g_k(\sigma) | \leqslant C_1 \exp[-| a_1\sigma |^{q'}] \qquad (k = 1, 2,..., r).$$

The function $V(\sigma)$ is bounded also for $t = t_1 + it_2$, and

$$\begin{aligned} \| e^{tP^*(\sigma)} \| &\leqslant \| \exp[t_1 P^*(\sigma) \| \cdot \| \exp[it_2 P^*(\sigma)] \| \\ &\leqslant \| \exp[| t_1 | P^*(\sigma) \operatorname{sgn} t_1] \| \cdot \| \exp[| t_2 | P^*(\sigma)\, i \operatorname{sgn} t_2] \| \\ &\leqslant C(1 + | \sigma |)^{2(m-1)p} \exp[C | t |(| \sigma |^{p_0} + 1)], \end{aligned}$$

Thus the integral admits the following estimate for large $| t |$

$$\left| \int_{-\infty}^{\infty} V(\sigma) \left\{ \sum_{k=0}^{r} t^k \exp[tP^*(\sigma)]\, P_k(\sigma) R_k(\partial/\partial\sigma)\psi(\sigma) \right\} d\sigma \right|$$

$$\leqslant B \sum_{k=0}^{r} | t |^k \int_{-\infty}^{\infty} \exp[C | t |(1 + | \sigma |^p)] \exp[-| a_1\sigma |^{q'}]\, d\sigma$$

$$\leqslant B \exp[C | t |]\, | t |^r \int_{-\infty}^{\infty} \exp[C | t |\, | \sigma |^p - | a_1\sigma |^{q'}]\, d\sigma. \tag{5}$$

The Young inequality (Chapter I, Section 3)

$$| uv | \leqslant \frac{u^\lambda}{\lambda} + \frac{v^{\lambda'}}{\lambda'} \qquad \left(\frac{1}{\lambda} + \frac{1}{\lambda'} = 1\right)$$

implies for the functions

$$u = C | t |, \qquad v = | \sigma |^{p_0};$$

the following estimate

$$\exp[C | t |\, | \sigma |^{p_0}] \leqslant \exp[C_1 | t |^\lambda + C_2 | \sigma |^{p_0\lambda'}].$$

We choose λ' so that

$$q' - \epsilon < p_0\lambda' < q',$$

where ϵ is a small quantity. We then have the following upper bound for (5):

$$B \exp[C \mid t \mid + C_1 \mid t \mid^{\lambda}] \mid t \mid^r \int_{-\infty}^{\infty} \exp[C_2 \mid \sigma \mid^{p_0\lambda'} - \mid a_1\sigma \mid^q] \, d\sigma,$$

This integral converges, so that the order of $F_\varphi(t)$ is not larger than λ.

Since, by definition, $(2p_0)' > q$, it follows that $2p_0 < q'$; furthermore

$$\lambda' > \frac{q' - \epsilon}{p_0} > \frac{2p_0 - \epsilon}{p_0} > 2,$$

hence

$$\lambda < 2.$$

The integral of the formal derivative with respect to t of the integrand is easily seen to be absolutely convergent. This implies that $F_\varphi(t)$ is indeed an entire function of the complex variable t, of order lower than 2.

We investigate the behavior of this function for real and purely imaginary values of t.

For real t, the estimate (6) in Section 7.1 yields

$$\mid F_\varphi(t) \mid = \mid (u(x, t), \varphi(x)) \mid \leqslant \int_{-\infty}^{\infty} \mid u(x, t)\varphi(x) \mid dx \leqslant C \exp[b \mid t \mid^{\gamma}].$$

In order to obtain an estimate of $F_\varphi(t)$ along the imaginary axis, we make use of the following theorem:

Let $P(\sigma)$ be an $m \times m$ matrix which for each $\sigma = (\sigma_1, \ldots, \sigma_n)$ has only real characteristic roots. Then for each real θ and in each bounded domain of the variable σ, the inequality

$$\| e^{i\theta P(\sigma)} \| \leqslant C \mid \theta \mid^{m-1} \tag{6}$$

holds.

The proof follows from the inequality

$$\| e^{tP} \| \leqslant e^{t\Lambda}(1 + 2t \| P \| + \cdots + (2t \| P \|)^{m-1}), \qquad \Lambda = \max_j \operatorname{Re} \lambda_j ,$$

which was proven in Section 6, by replacing t by $\mid \theta \mid$ (θ is real) and $P(\sigma)$ by $iP(\sigma) \operatorname{sgn} \theta$. The matrix $iP(\sigma) \operatorname{sgn} \theta$ has only imaginary zeros λ_j.

We apply the estimate (6) to the equality (4) for purely imaginary $t = i\theta$. Since the characteristic roots of the matrix $P^*(\sigma)$ are real at the same time as those of the matrix $P(\sigma)$, it follows that $F_\varphi(t)$ does not grow along the imaginary axis faster than a polynomial of degree $r_0 = r + m - 1$.

Thus the function $F_\varphi(t)$ does not increase faster than $\exp(b|t|^\gamma)$ along all four coordinate semi-axes of the t-plane. Since, as was proved, $F_\varphi(t)$ is also of order smaller than 2, the remark after the Phragmén-Lindelöf theorem (Section 7, Chapter IV, Volume 2) implies that it is of order (lower than) γ in each of the four quandrants of the t-plane. But then $F_\varphi(t)$ is an entire function of order $\leqslant \gamma$. It does not increase faster than a polynomial along the imaginary axis, i.e., along the sides of an angle with opening $\theta < \pi/\gamma$, and therefore, on that basis, is itself a polynomial in t of degree not higher than r_0.

We now show that the function $u(x, t)$ is itself a polynomial in t of degree not higher than r_0. In order to do this, we establish two simple lemmas.

Lemma 1. *If the sequence of polynomials of degree r, $R_\nu(t)$ ($\nu = 1, 2, \ldots$; $-\infty < t < \infty$) converges for each t to a function $R(t)$, then $R(t)$ is itself a polynomial of degree not higher than r.*

Proof. Let

$$R_\nu(t) = a_{0\nu} + a_{1\nu}t + \cdots + a_{r\nu}t^r.$$

For $t = 0$ the a_0 have the limit $a_{0\nu}$ for $\nu \to \infty$. Consequently, the polynomials

$$a_{1\nu} + \cdots + a_{r\nu}t^{r-1}$$

converge for all t (to a limit equal to $(R(t) - a_0)/t$ for $t = 0$). It follows that the $a_{1\nu}$ have a limit a_1, and each sequence $a_{i\nu}$ has a limit a_i. Defining

$$R_0(t) = a_0 + a_1 t + \cdots + a_r t^r,$$

we will obviously have $R_\nu(t) \to R_0(t)$ for all t. Thus $R(t) \equiv R_0(t)$, i.e., $R(t)$ is indeed a polynomial of degree $\leqslant r$.

Lemma 2. *Let $u(x, t)$ be a vector function satisfying the inequalities in Section 7.1; then for any real x_0, the function $u(x - x_0, t)$ also satisfies the same inequalities, possibly with different constants C, a_0, and b_0.*

Proof. It is geometrically obvious that for any $b > 0$ there exists a $c > 0$, such that for any real ξ the following inequality is satisfied:

$$b\,|\,\xi - 1\,|^\gamma \leqslant b\,|\,\xi\,|^\gamma + c.$$

Replacing ξ by x_j/x_j^0 we find:

$$b\,|\,x_j - x_j^0\,|^\gamma \leqslant b\,|\,x_j\,|^\gamma + c\,|\,x_j^0\,|^\gamma,$$

whence, upon summation over j, it follows

$$b \sum_{t=1}^{n} | x_j - x_j^0 |^\gamma \leqslant b \sum_{t=1}^{n} | x_j |^\gamma + c \sum_{t=1}^{n} | x_j^0 |^\gamma,$$

which yields the required inequality.

The fact that the $u_j(x, t)$ are polynomials in t can be proved as follows. Consider an even, nonnegative nonvanishing function $\psi_0(\sigma) \in W_q^{q'}$. The convolution $\psi_1(\sigma) = \int \psi_0(\sigma - \xi)\, \psi_0(\xi)\, d\xi$ also belongs to $W_q^{q'}$ (cf. the discussion at the and of Section 6, Chapter IV, Volume 2). The Fourier transforms $\varphi_0(x)$ and $\varphi_0(x)$ of $\psi_0(x)$ and $\psi_1(x)$, respectively, belong to the space $W_q{}^q$, since $W_q{}^q = S_{1/q}^{1-1/q}$. Here $\varphi_0(x)$ is a real function and $\varphi_1(x) = \varphi_0{}^2(x)$ is nonnegative, with

$$\varphi_1(0) = \varphi_0{}^2(0) = \left(\int \psi_0(\sigma)\, d\sigma\right)^2 > 0.$$

Consider the functions

$$\varphi_\nu(x, x_0) = \frac{\varphi_1[\nu(x - x_0)]}{\int \varphi_1[\nu(x - x_0)]\, dx} \qquad (\nu = 2, 3, \ldots).$$

which also belong to $W_q{}^q$ and in addition have the properties:

(a) $\varphi_\nu(x, x_0) \geqslant 0,$

(b) $\int \varphi_\nu(x, x_0)\, dx = 1,$

and, since

$$\int \varphi_1[\nu(x - x_0)]\, dx = \frac{1}{\nu} \int \varphi_1(x)\, dx = \frac{1}{C_1 \nu},$$

we also have

(c) $| \varphi_\nu(x, 0) | \leqslant C_1 \nu \, | \varphi_1(\nu x) | \leqslant C_2 \nu \exp[-a(\nu x)^{1/\alpha}].$

For any fixed t we have

$$\lim_{\nu \to \infty} \int u(x, t) \varphi_\nu(x, x_0)\, dx = u(x_0, t). \tag{7}$$

In order to prove this, we first consider the case $x_0 = 0$. Then

$$\left| u(0, t) - \int u(x, t) \varphi_\nu(x, 0)\, dx \right| \leqslant \int | u(0, t) - u(x, t) | \, \varphi_\nu(x, 0)\, dx.$$

Since $u(x, t)$ is continuous in x for $x = 0$, we have for an arbitrary $\epsilon > 0$ and a sufficiently small $\delta > 0$

$$\int_{|x|<\delta} |u(0, t) - u(x, t)| \varphi_\nu(x, 0)\, dx \leqslant \epsilon \int_{|x|\leqslant\delta} \varphi_\nu(x, 0)\, dx \leqslant \epsilon. \tag{8}$$

The integral over the complement of the ball $|x| < \delta$ can be estimated as follows:

$$\begin{aligned}\int_{|x|\geqslant\delta} &|u(0, t) - u(x, t)| \varphi_\nu(x, 0)\, dx \\ &\leqslant 2C_3 \int_{|x|\geqslant\delta} \exp[b_0 |x|^\gamma \nu e^{-a(\nu x)^{1/\alpha}}]\, dx \\ &\leqslant C_4 \int_{|\xi|\geqslant\nu\delta} \exp[b_0 |\xi/\nu|^\gamma - a|\xi|^{1/\alpha}]\, d\xi \\ &\leqslant C_4 \int_{|\xi|\geqslant\nu\delta} \exp[b_1 |\xi|^\gamma - a_1 |\xi|^{1/\alpha}]\, d\xi \leqslant \epsilon \end{aligned} \tag{9}$$

for sufficiently large ν.

The inequalities (8) and (9) imply (7) for $x_0 = 0$. For an arbitrary value of x_0 we make use of Lemma 2. Then:

$$\begin{aligned}\int u(x, t)\varphi_\nu(x, x_0)\, dx &= \int u(x, t)\varphi_\nu(x - x_0, 0)\, dx \\ &= \int u(\xi + x_0, t)\varphi_\nu(\xi, 0)\, d\xi,\end{aligned}$$

which, on the basis of Lemma 2, and of that just proved, converges to $u(x_0, t)$. Thus the limit relation (7) is proved.

As we have seen above, the functions $U_\nu(t) = \int u(x, t)\, \varphi_\nu(x, x_0)\, dx$ are polynomials in t of degree not larger than r_0. Making use of Lemma 1, it follows that $u(x_0, t)$ is also a polynomial in t, of degree not higher than r_0:

$$u(x_0, t) = \sum_{k=0}^{r_0} U_k(x_0)t^k,$$

or, also

$$u(x, t) = \sum_{k=0}^{r_0} U_k(x)t^k. \tag{10}$$

As a solution of the system (5), Section 7.1, the vector function $u(x, t)$ is differentiable with respect to x, or, more precisely, is in the domain of the differential operator $P(i\, \partial/\partial x)$. We must show that each coefficient $U_k(x)$, $(k = 0, 1,..., r_0)$ possesses the same property. This is established with the help of the following lemma.

Lemma 3. *The order of differentiability of the coefficients $U_k(x)$ with respect to x is the same as the order of differentiability of the vector-function $u(x, t) = \sum_{k=0}^{r_0} U_k(x)\, t^k$ with respect to x, for any fixed t.*

Proof. Setting successively $t = 0, 1, 2, \ldots, r_0$, we obtain for $U_k(x)$ the system of equations

$$\begin{aligned} u(x, 0) &= U_0(x), \\ u(x, 1) &= U_0(x) + U_1(x) + \cdots + U_{r_0}(x), \\ u(x, 2) &= U_0(x) + 2U_1(x) + \cdots + 2^{r_0} U_{r_0}(x), \\ &\cdots \\ u(x, r_0) &= U_0(x) + r_0 U_1(x) + \cdots + r_0^{r_0} U_{r_0}(x). \end{aligned}$$

Since the determinant of this system is of the Vandermonde type, and hence nonvanishing, we can solve with respect to $U_0(x), U_1(x), \ldots, U_{r_0}(x)$. The solutions will be linear (with constant coefficients) in $u(x, 0)$, $u(x, 1), \ldots, u(x, r_0)$ and hence will be differentiable exactly as many times as the latter ones. Thus the lemma is proved.

Substituting (8) into the equations (5), Section 7.1 and differentiating term by term with respect to x (which is legitimate, according to lemma 3) we obtain

$$\sum_{k=0}^{r_0} k U_k(x) t^{k-1} = \sum_{k=0}^{r_0} t^k P\left(i \frac{\partial}{\partial x}\right) U_k(x),$$

whence

$$k U_k(x) = P\left(i \frac{\partial}{\partial x}\right) U_{k-1}(x);$$

in particular

$$P\left(i \frac{\partial}{\partial x}\right) U_{r_0}(x) = 0,$$

$$P\left(i \frac{\partial}{\partial x}\right) U_{r_0 - 1}(x) = r_0 U_{r_0}(x), \ldots$$

Thus the theorem is completely proved.

Setting $t = 0$ in Equation (8), Section 7.1, we have $u(x, 0) = U_0(x)$ (in vector notation) and (9) implies $P^r(i\, \partial/\partial x)\, u(x, 0) = 0$. Thus the Cauchy data for the solution under consideration are themselves solutions of a system of equations in the variables x. Recalling the results of Section 2.4.2, Chapter III, Volume 2, we can strengthen the formulation of the theorem:

This implies that the norm of the matrix $\widetilde{f(s)} = \| \widetilde{f_{jk}}(s) \|$ satisfies the inequality

$$\| \widetilde{f(s)} \| \leqslant C_2[(1 + |\sigma|)^{hp} + e^{b_0 r|\tau|}].$$

and therefore

$$\| Q(s, t_0, t) \| \leqslant \exp[(t - t) \| \widetilde{f(s)} \|] \leqslant \exp[C_2(t - t_0)(1 + |\sigma|)^{hp}]$$
$$\times \exp[C_2(t - t_0)e^{b_0 r|\tau|}].$$

As a test function space we choose now one of the (vector-) spaces $W_{M,a}^{\Omega,b}$ (Section 1, Chapter I). This space consists of vectors, the components of which are entire analytic functions $\varphi(x + iy)$ satisfying the inequalities

$$|\varphi(x + iy)| \leqslant C \exp[-M[(a - \delta)x] + \Omega[(b + \rho)y]]$$

where $M(x)$ and $\Omega(y)$ are convex functions. In Section 3, Chapter I it was shown that the Fourier dual space $\tilde{\Phi} = \Psi$ is of the same type

$$\widetilde{W_{M,a}^{\Omega,b}} = W_{M_1,1/b}^{\Omega_1,1/a},$$

where the functions M_1 and Ω_1 are the Young duals of Ω and M, respectively. We set $M_1(\sigma) = \sigma^q/q$ where q is an arbitrary number larger than hp and $\Omega_1(\tau) = e^\tau - \tau - 1$. Then, according to Section 2, Chapter I, the functions making up the matrix $Q(s, t_0, t)$ are multipliers in the space $W_{M_1,1/b}^{\Omega_1,1/a}$ and map this space into the space $W_{M_1,1/b}^{\Omega_1,1/a+b_0 r}$. This, in turn, shows that the operators which make up the matrix $Q(i\,\partial/\partial x, t_0, t)$ are defined and bounded in the space $\Phi \equiv W_{M,a}^{\Omega,b}$ and map it into the space $\Phi_1 = W_{M,a'}^{\Omega,b}$, $1/a' = (1/a) + b_0 r$. The function $M(x)$ which is Young-dual to $\Omega_1(\tau)$ is

$$M(x) = (x + 1)\ln(x + 1) - x$$

and is equivalent to the function $x \ln x$.

Thus the components of the vector belonging to the space Φ decrease along the x-axis according to the inequality

$$|\varphi(x)| \leqslant C_\delta \exp[-(a' - \delta)|x| \ln |x|], \quad \frac{1}{a'} = \frac{1}{a} + b_0 r.$$

The number a can be chosen arbitrarily large, the number r may be as close as convenient to 1, therefore the coefficient inside the round brackets in the exponent can be made arbitrarily close to $1/b_0$. We denote this coefficient by $(1/b_0) - \delta'$.

As space E we select the space of functions $\varphi(x)$ with the norm defined by

$$\| \varphi \|_E = \int_{-\infty}^{\infty} \exp\Big(\frac{1}{b_0 + \delta} \, | x | \ln | x |\Big) \| \varphi(x) \| \, dx$$

with arbitrary $\delta > 0$. Then for sufficiently small δ' the spaces Φ, Φ_1 will be completely contained in the space E.

Applying the theorem of Section 4, we obtain the following result.

If the solution of the system

$$\frac{\partial u_j(x, t)}{\partial t} = - \sum_{k=1}^{m} f_{kj} * u_k(x, t) \qquad (j = 1, 2, \ldots, m) \tag{7}$$

is a set of functionals defined on the space Φ, which for all t $(0 \leqslant t \leqslant T)$ belong to the space E', this solution is uniquely determined by its Cauchy data

$$u_j(x, 0) = u_j(x) \in E'. \tag{8}$$

In the same manner as in Section 3, it can be shown that any classical solution of the system (7), i.e., a set of functions $u_j(x, t)$ satisfying the system and the inequalities

$$| u_j(x, t) | \leqslant C \exp\Big[\Big(\frac{1}{b_0} - \delta'\Big) | x | \ln | x |\Big],$$

is also a generalized solution of the system in E'. Therefore *the class of functions $u(x)$ satisfying the inequality*

$$| u(x) | \leqslant C \exp\Big[\Big(\frac{1}{b_0} - \delta'\Big) | x | \ln | x |\Big], \tag{9}$$

for any fixed δ' *is a uniqueness class* of solutions of the Cauchy problem for the system of convolution equations (7).

In particular, for a difference system corresponding to convolutions with the functionals $f_{jk} = \delta(x - x_{jk})$, the number b_0, defining the maximal length of the intervals which support the functionals f_{jk}, coincides with the number $H = \max | x_{jk} |$ (the maximal translation). Therefore, for a system of difference equations, the coefficient $1/b_0 - \delta'$ in the exponent of (9), which determines the uniqueness class for the solutions of the Cauchy problem, can be replaced by $1/H - \delta$. The same is also true for a difference-differential-system, defined by convolutions with $\delta^{(q)}(x - x_{jk})$.

Under certain additional restrictions, one may consider convolution equations with functionals f_{jk} having noncompact support. For instance, consider a functional f_{jk} which is of the type of an entire analytic function belonging to the space S_α^β, $\alpha + \beta = 1$. Then $\widetilde{f_{jk}}(s)$ is an entire analytic function belonging to the space S_β^α and consequently satisfying the inequality

$$|\widetilde{f_{jk}}(\sigma + i\tau) \leqslant C \exp[-a_0 | \sigma |^{1/\beta} + b_0 | \tau |^{1/(1-\alpha)}] \leqslant C \exp[b_0 | \tau |^{1/(1-\alpha)}].$$

By a similar argument as above, we find that the norm of the matrix satisfies the inequality

$$\| Q(s, t_0, t) \| \leqslant C_1 \exp[C(t - t_0) \exp[b_0 | \tau |^{1/(1-\alpha)}].$$

Such a matrix $Q(s, t_0, t)$ defines a multiplier in the space W^Ω where $\Omega(\tau) = e^{b|\tau|^{1/(1-\alpha)}}$ The Fourier-dual of this space is W_M with $M(x)$ equivalent to $| x | \ln^{1-\alpha} x$. Therefore, using the same reasoning as above, we find that the *uniqueness class of the solution of the corresponding Cauchy problem is the class of functions $u(x)$ satisfying the inequality*

$$| u(x) | \leqslant C \exp[| x | \ln^{1-\alpha} | x |]$$

APPENDIX 2

Equations with Coefficients Which Depend on x

A2.1. The General Scheme

Consider a continuous family of normed spaces $(a < \lambda < b < \infty)$ such that $\Phi_\lambda \supset \Phi_\mu$ for $\lambda < \mu$; the intersection of this family is $\Phi = \cap \Phi_\lambda$. Consider further a linear operator A mapping each of the spaces Φ_μ into any of the wider spaces $\Phi_\lambda (\lambda < \mu)$ (one could also denote such an operator by $A_\mu^\lambda : \Phi_\mu \to \Phi_\lambda$). We assume that the norm of A is

$$\| A \|_\mu^\lambda \leqslant \frac{C}{\mu - \lambda}.$$

We show that the operator $Q = e^{(t-t_0)A}$ is bounded for sufficiently small $t - t_0$, and maps each space Φ_μ into any of the larger spaces Φ_λ. For the proof, we partition the interval $\lambda \leqslant \xi \leqslant \mu$ into n equal parts of length $(\mu - \lambda) n$ by menas of the points $\xi_0 = \lambda, \xi_1, \xi_2, \ldots, \xi_n = \mu$.

The operator A^n can be represented as a product of n operators A, each mapping the space $\Phi_{\xi_{j+1}}$ into Φ_{ξ_j}. In particular, we obtain

$$\| A^n \|_\mu^\lambda \leqslant \prod_{i=1}^{n-1} \| A_{\xi_{j+1}}^{\xi_j} \| \leqslant \left(\frac{Cn}{\mu - \lambda}\right)^n = C_1^n n^n.$$

Therefore the series

$$\sum_{n=0}^{\infty} \frac{(t-t_0)^n}{n!} \| A^n \|_\mu^\lambda \leqslant \sum_{n=0}^{\infty} C_1^n \frac{n^n}{n!} (t - t_0)^n$$

converges for $t - t_0 < 1/(C_1 e)$. This means that for the indicated values of t, there exists the operator $Q = e^{(t-t_0)A}$ mapping the space Φ_μ into Φ_λ, as asserted.

This operator $Q \equiv Q_{t_0}^t$ allows one to construct a solution of the Cauchy problem for the equation

$$\frac{d\varphi(t)}{dt} = A\varphi(t)$$

with Cauchy data $\varphi(t_0) = \varphi_0$. The solution $\varphi(t) = Q_{t_0}^t \varphi(t_0)$ belongs to the space Φ_λ if $\varphi(t_0) \in \Phi_\mu$, $\lambda < \mu$, and $t - t_0$ is sufficiently small.

A2.2. Systems with Convolution Operators

Let us consider the system of convolution equations of the form

$$\frac{\partial \varphi_j(x, t)}{\partial t} = \sum_{k=1}^{m} a_{jk}(x) \int_{-\infty}^{\infty} \varphi_k(x - \xi, t)\, d\rho_{jk}(\xi), \tag{1}$$

where the functions $\rho_{jk}(\xi)$ are of bounded variation in $(-\infty, \infty)$ and

$$\int_{-\infty}^{\infty} e^{\mu|x|} \,|\, d\rho_{jk}(\xi) \,| = C_\mu < \infty \tag{2}$$

at least for some interval of values of $0 \leqslant \mu_1 < \mu < \mu_2$.

We assume that the functions $a_{jk}(x)$ satisfy the inequalities

$$|\, a_{jk}(x) \,| \leqslant a \,|\, x \,| + b.$$

We denote by Φ_λ $(\mu_1 \leqslant \lambda \leqslant \mu_2)$ the normed space of functions $\varphi(x)$ with the norm

$$\| \varphi \|_\lambda = \sup_x |\, \varphi(x) \,|\, e^{\lambda|x|}.$$

(In the vector case we have to substitute for $|\varphi(x)|$ the Euclidean norm of the vector $\varphi(x)$). Obviously, for $\lambda < \mu$ we have $\Phi_\lambda \supset \Phi_\mu$.

We show that the operators of the right-hand side of the system (1) satisfy the conditions of Section 2A.1. If $\varphi(x)$ belongs to Φ and

$$\psi(x) = \int_{-\infty}^{\infty} \varphi(x - \xi)\, d\rho(\xi), \tag{3}$$

where $\rho(\xi)$ is a function of bounded variation, satisfying an inequality of type (2), we have

$$\begin{aligned}\| \psi(x) \|_\mu &= \sup | \psi(x) | e^{\mu|x|} = \sup e^{\mu|x|} \left| \int_{-\infty}^{\infty} \varphi(x - \xi)\, d\rho(\xi) \right| \\ &\leqslant \sup_x e^{\mu|x|} \int_{-\infty}^{\infty} \| \varphi \|_\mu e^{-\mu|x-\xi|} \, | d\rho(\xi) | \\ &= \| \varphi \|_\mu \sup_x \int_{-\infty}^{\infty} e^{+\mu|x|-\mu|x-\xi|} \, | d\rho(\xi) | \\ &\leqslant \| \varphi \|_\mu \sup_x \int_{-\infty}^{\infty} e^{\mu|\xi|} \, | d\rho(\xi) | = C_\mu \| \varphi \|_\mu .\end{aligned}$$

Thus the operator (3) maps the space Φ_μ into itself and has a norm which does not exceed C_μ.

Let $\psi(x) \in \Phi_\mu$ and

$$\psi_a(x) = a(x)\psi(x),$$

where $a(x)$ satisfies the inequality

$$| a(x) | \leqslant a \, | x | + b. \tag{4}$$

Obviously, the function $\psi_a(x)$ can belong to any space Φ_λ for $\lambda < \mu$ (but in general does not belong to Φ_μ itself). We have:

$$\begin{aligned}\| \psi_a(x) \|_\lambda &= \sup | a(x)\psi(x) | e^{\lambda|x|} \\ &\leqslant \sup | a(x) | e^{(\lambda-\mu)|x|} \cdot \sup | \psi(x) | e^{\mu|x|}.\end{aligned}$$

We estimate the first factor in the right-hand side. According to (4)

$$| a(x) | e^{(\lambda-\mu)|x|} \leqslant a \, | x | \, e^{(\lambda-\mu)|x|} + b e^{(\lambda-\mu)|x|}.$$

By means of differentiation it is easy to see that the maximum of the first term does not exceed $a/(e(\mu - \lambda))$. It is obvious that the second term does not exceed b.

Thus

$$\| \psi_a(x) \|_\lambda \leqslant \left[\frac{a}{e(\mu-\lambda)} + b\right] \| \psi(x) \|_\mu ,$$

which implies that the operation of multiplication by $a(x)$, considered as an operator from Φ_μ into Φ_λ, $\lambda < \mu$, has a norm which does not exceed

$$\frac{a}{e(\mu-\lambda)} + b \leqslant \frac{a_1}{\mu-\lambda} \qquad \text{for } \mu - \lambda < C.$$

Consequently each element of the matrix of the system (1) is an operator mapping the space Φ_μ into Φ_λ, $\lambda < \mu$, with a norm bounded by $a_1(\mu - \lambda)$.

The system (1) can be rewritten in vector form

$$\frac{d\varphi(t)}{dt} = A\varphi(t),$$

where $\varphi(t)$ is a vector with coordinates $\varphi_j(x, t)$. A is the operator in the m-dimensional space, defined by the matrix of the system (1). This operator obviously maps the vector space Φ_μ into any Φ_λ $\lambda < \mu$ and its norm does not exceed $C/(\mu - \lambda)$.

According to the results of Section A.2.1, the Cauchy problem for the system (1) with initial data belonging to the space Φ_μ always has a solution

$$\varphi(t) = e^{(t-t_0)A}\varphi(t_0)$$

for any space $\Phi_\lambda \supset \Phi_\mu$. On the basis of the fundamental theorem in Section 2, the Cauchy problem for the inverse-conjugate system

$$\frac{\partial u_j(x, t)}{\partial t} = -\sum_{k=1}^{m} \bar{a}_{kj}(x) \int u_k(x-\xi)\, d\bar{\rho}_{jk}(\xi) \tag{5}$$

can only have a unique solution in the space which is dual to any Φ_λ. In particular, the uniqueness class for the Cauchy problem for the system (5) is the totality of functions $f(x)$ which satisfy the inequality

$$| f(x) | \leqslant Ce^{\lambda|x|} \tag{6}$$

for any fixed λ, $\mu_1 < \lambda < \mu_2$.

If the $\rho_{jk}(\xi)$ are step functions, the system (1) becomes a system of pure difference equations. Thus, for systems of difference equations (1) with coefficients which do not increase faster than the first power of the variable, the class of functions $f(x)$ defined by the inequality (6) is the uniqueness class for solutions of the Cauchy problem.

We obtain the same result if we replace the test function space Φ_λ consisting of functions of a real variable $\varphi(x)$, exponentially decreasing along the x-axis, by a space Φ_λ consisting of entire analytic functions of order $h > 1$, with the norms defined by

$$\| \varphi \|_\lambda = \sup | \varphi(x + iy) | \exp[\lambda | x | + \lambda' | y |^h] \qquad \left(\lambda' = -\frac{1}{\lambda}\right) \tag{7}$$

(in the vector case we have again to replace $| \varphi(x + iy)|$ by the Euclidean norm of the vector φ). As coefficients $a(x)$ we can take linear functions of $z = x + iy$.

The proof proceeds along the same lines as before. One must only majorize the quantity

$$A = \sup | y | \exp\left[\left(\frac{1}{\mu} - \frac{1}{\lambda}\right) | y |^h\right].$$

This majorization is easily achieved by means of differentiation, namely

$$A \leqslant \frac{C}{(\mu - \lambda)^{1/h}} \leqslant \frac{C_1}{\mu - \lambda} \qquad \text{for } \mu - \lambda \leqslant C_0 ,$$

as required. This second form will be used in the next section.

A2.3. Kovalevskaya Systems

A system partial of differential equations of the form

$$\frac{\partial u_j(x, t)}{\partial t} = \sum_{k=1}^{m} \left\{ i a_{jk}(x) \frac{\partial u_k(x, t)}{\partial x} + b_{jk}(x) u_k(x, t) \right\} \tag{1}$$

is called a Kovalevskaya[13] system. We show that *if the functions* $a_{jk}(x)$ *and* $b_{jk}(x)$ *are Fourier transforms of exponentially decreasing measures, i.e.*

$$a_{jk}(x) = \int_{-\infty}^{\infty} e^{i\sigma x}\, d\rho_{jk}(\sigma), \qquad b_{jk}(x) = \int_{-\infty}^{\infty} e^{i\sigma x}\, d\rho^*_{jk}(\sigma), \tag{2}$$

with

$$\int_{-\infty}^{\infty} e^{\mu|\sigma|}\, | d\rho_{jk}(\sigma) | < \infty, \qquad \int_{-\infty}^{\infty} e^{\mu|\sigma|}\, | d\rho^*_{jk}(\sigma) | < \infty \qquad (| \mu | \leqslant \mu_1), \tag{3}$$

[13] *Translator's note:* We have used the russian feminine Kovalevskaya in place of the transcription Kowalewsky often used in Western literature.

then the class of functions $f(x)$ satisfying the inequality

$$| f(x) | \leqslant Ce^{|x|^p},$$

for any fixed p, is a uniqueness class for solutions of the Cauchy problem.

Remark. In order that the function $a_{jk}(x)$ and $b_{jk}(x)$ be representable in the form (2)–(3) it is *necessary* that they be analytically continuable into the strip $| \operatorname{Im} z | \leqslant \mu_1$ and remain bounded within this strip. A *sufficient* condition for this representability is that the functions obtained by analytic continuation possess a majorant which is absolutely integrable for $-\infty < x < \infty$. In this case the Stieltjes integral reduces to an (improper) Riemann integral.

Proof. We choose as test function space Φ the space of functions $\varphi(x)$ which can be continued analytically into a strip $| \operatorname{Im} z | \leqslant \lambda$ (λ depends on $\varphi(x)$) and satisfy the estimates

$$| \varphi(x + iy) | \leqslant C \exp[a | x |^p] \qquad (p > 1),$$

where the constant C is bounded in the strip $| y | \leqslant \lambda$. This space coincides with $S^1_{1/p}$ (Section 2, Chapter IV, Volume 2).

It was shown in Volume 2, Chapter IV that the Fourier dual space $\Psi = \tilde{\Phi} = S_1^{1/p}$ consists of entire analytic functions $\psi(\sigma + i\tau)$ satisfying

$$| \psi(\sigma + i\tau) | \leqslant C \exp[-a' | \sigma | + b' | \tau |^{p'}],$$

with $1/p + 1/p' = 1$ and a' is a constant depending on ψ.

The inverse-adjoint system of (1) (in the sense of Section 3.2),

$$\frac{\partial \varphi_j(x, t)}{\partial t} = -\sum_{k=1}^{m} \left\{ i \frac{\partial}{\partial x} [\bar{a}_{kj}(x) \varphi_k(x, t)] + \bar{b}_{kj}(x) \varphi_k(x, t) \right\}. \tag{1'}$$

considered as a system with unknown test functions $\varphi_k(x, t)$, can be subjected to Fourier transformation.

The products $\bar{a}_{kj}(k)\, \varphi_k(x, t)$ are transformed into the convolutions $\tilde{\bar{a}}_{kj}(\sigma) * \psi_k(\sigma, t)$. Since the Fourier transform of a_{kj} is by definition a measure, we have

$$\tilde{\bar{a}}_{kj}(\sigma) * \psi_k(\sigma, t) = \int_{-\infty}^{\infty} \psi_k(\sigma - \xi, t)\, d\tau_{kj}(\xi),$$

with the function $\tau_{kj}(\sigma)$ of bounded variation for $-\infty < \sigma < \infty$.

Thus (1′) becomes

$$\frac{\partial \psi_j(\sigma, t)}{\partial t} = \sum_{k=1}^{m} \left\{ \sigma \int_{-\infty}^{\infty} \psi_k(\sigma - \xi, t)\, d\rho_{jk}(\xi) + \int_{-\infty}^{\infty} \psi_k(\sigma - \xi, t)\, d\hat{\rho}_{jk}(\xi) \right\}. \tag{4}$$

It was shown in Section A2.2 of this appendix that for such a convolution system the operator $\underset{\sim}{Q}_{t_0}^{t}$ which solves the Cauchy problem exists.

According to the general theorem of Section A2.2, the Fourier-transformed $\tilde{\underset{\sim}{Q}}{}_t^{t_0} = G_{t_0}^{t}$ solves the corresponding problem in the space Φ.

Consequently, the Cauchy problem for the system

$$\frac{\partial u_j(x, t)}{\partial t} = \sum_{k=1}^{m} \left\{ a_{jk}(x) i \frac{\partial}{\partial x} u_k(x, t) + b_{jk}(x) u_k(x, t) \right\} \tag{1}$$

has a unique solution in the space Φ'. It follows, in particular, that the functions $f(x)$ which satisfy an inequality

$$| f(x) | C \exp[| x |^{p_1}]$$

for $p_1 < p$, form a uniqueness class for the Cauchy problem of Eq. (1), since these functions define continuous linear functionals on the space Φ. It is easy to see that p_1 as well as p are arbitrary. This completes the proof of the theorem.

APPENDIX 3

Systems with Elliptic Operators

A differential operator $L(D) = \sum_{|q| \leqslant p_1} a_q(x) D^q$ is said to be *elliptic* if the sum of its highest order terms $L_0(D) = \sum_{|q|=p} a_q(x) D^q$ has the following property: For any fixed x, the function $L_0(\sigma) = \sum_{|q|=p} a_q(x) \sigma^q$ is nonnegative for all real σ, and vanishes only for $\sigma = 0$. The coefficients $a_q(x)$ are assumed to be continuous everywhere and bounded, together with their first-order derivatives.

Consider a system of the form

$$\frac{\partial u_j(x, t)}{\partial t} = \sum_{k=1}^{m} P_{jk}(L, t) u_k(x, t) \qquad (j = 1, 2, \ldots, m), \tag{1}$$

where P_{jk} are polynomials with coefficients which are functions of t for

$0 \leqslant t \leqslant T$. We are interested in solutions satisfying the initial conditions

$$u_j(x, 0) = u_j(x) \qquad (j = 1, 2, \ldots, m). \tag{2}$$

Theorem. *If the resolvent matrix $Q(s, t_0, t)$ of the inverse-adjoint system of* (1)

$$\frac{\partial \psi_j(s, t)}{\partial t} = \sum_{k=1}^{m} \tilde{P}_{jk}(s, t)\psi_k(s, t), \tag{3}$$

admits in the s-plane ($s = \sigma + i\tau$) an estimate

$$\| Q(s, t_0, t) \| \leqslant C_1 \exp[b(t - t_0) \mid s \mid], \tag{4}$$

then the functions $f(x)$ which satisfy the inequality

$$\mid f(x) \mid \leqslant C \exp[b_1 \mid x \mid^{p_1'}] \qquad \left(\frac{1}{p_1} + \frac{1}{p_1'} = 1\right), \tag{5}$$

form a uniqueness class for the Cauchy problem (1), (2).

If the coefficients of the polynomials $\tilde{P}_{jk}(s, t) = \tilde{P}_{jk}(s)$ are constants, then in order that (4) be true it is sufficient that the characteristic roots (s) of the matrix $\tilde{P}(s)$ satisfy the inequality

$$\Lambda(s) = \max_j \operatorname{Re} \lambda_j(s) \leqslant C(1 + \mid s \mid). \tag{6}$$

Before giving the proof, we note some of the properties of the solution of the Cauchy problem for equations of the form

$$\partial u / \partial t = Lu \tag{7}$$

where L is an elliptic operator. We shall make use of the following results due to S. D. Eĭdelman.

Equation (7) *always admits a solution if the initial function $u(x, 0)$ satisfies the inequality*

$$\mid u(x, 0) \mid \leqslant C_1 \exp[-a \mid x \mid^{p_1'}] \qquad \left(\frac{1}{p_1} + \frac{1}{p_1'} = 1\right) \tag{8}$$

here C_1 and a are constants. The solution $u(x, t)$ satisfies, for each $t \geqslant 0$, the inequality

$$\mid u(x, t) \mid \leqslant C_1' \exp[-a' \mid x \mid^{p_1'}] \tag{9}$$

For $t > 0$, $u(x, t)$ is an infinitely differentiable function of t, and its derivatives with respect to t satisfy the inequalities

$$\left| \frac{\partial^q u(x, t)}{\partial t^q} \right| \leqslant CB^q q! \exp[-a' \mid x \mid^{p_1'}] \qquad (q = 0, 1, 2, \ldots). \tag{10}$$

Differentiating both sides of (7) with respect to t, we have

$$\frac{\partial}{\partial t} \frac{\partial u(x, t)}{\partial t} = L \frac{\partial u(x, t)}{\partial t},$$

i.e., the function $\partial u(x, t)/\partial t$ is also a solution of the equation (7). Since $L(\partial u(x, t)/\partial t) = L(Lu)$, the operator L can be applied to u twice. Repeating this argument, one can see that L can be applied to $u(x, t)$ arbitrarily often. Due to the inequality (10) we have

$$| L^q u(x, t) | \leqslant CB^q q! \exp[-a' \mid x \mid^{p_1'}]$$

We can now prove the theorem. As test function space Φ we choose the space of all functions $\varphi(x)$ defined for real x which can be subjected an arbitrary number of times to the operator L and satisfy the inequalities

$$\| L^q \varphi(x) \| \leqslant B^q q! \exp[-a \mid x \mid^{p_1'}].$$

It follows from the results quoted above, that such functions exist indeed. For example, solutions of Eq. (7) with initial conditions satisfying (8) satisfy the requirements for $t > 0$. One can even show that *the space Φ is sufficiently rich* in functions, in the following sense: if for some locally integrable function $f(x)$ the integral

$$\int_{-\infty}^{\infty} f(x)\varphi(x)\, dx$$

exists for all $\varphi \in \Phi$, and vanishes for every $\varphi \in \Phi$, then $f(x) \equiv 0$ almost everywhere. For the proof we consider a solution $u(x, t)$ of the Cauchy problem (7), (8) corresponding to an initial function $u_0(x)$ of compact support. Since for each $t > 0$ $u(x, t) \in \Phi$, we have by assumption

$$\int_{-\infty}^{\infty} f(x)u(x, t)\, dx = 0. \tag{11}$$

The function $u(x, t)$ satisfies the inequality (9) with C_1' independent of t. If $f(x)$ in turn satisfies the inequality

$$\| f(x) \| \leqslant A_1 \exp[A \mid x \mid^{p_1'}], \qquad A < a',$$

one can go to the limit $t \to 0$ in the integral (11) and we obtain:

$$\int_{-\infty}^{\infty} f(x)u_0(x)\,dx = 0.$$

Since $u_0(x)$ is an arbitrary function of compact support, it follows indeed that $f(x) = 0$ almost everywhere.

The test function space constructed by us corresponds to the space $S_{\alpha,A}^{1,B}$ with $\alpha = 1/p_1'$, the operator $\partial/\partial x$ being replaced by L. We therefore denote Φ by $LS_{\alpha,A}^{1,B}$. The principal theorem of Section 5, Chapter IV, Volume 2 is valid in $LS_{\alpha,A}^{1,B}$. We reformulate this theorem as follows:

Let $f(s)$ be an entire function of order 1 and type smaller than $1/(Be^2)$; then the operator $f(L)$ is defined and bounded on the space $\Phi = LS_{\alpha,A}^{1,B}$ and maps this space into the space $\Phi_1 = LS_{\alpha,A}^{1,Be}$.

The proof can be taken over literally from Volume 2, by replacing $\partial/\partial x$ by L and β by 1.

By assumption the resolvent matrix $Q(s, t_0, t)$ of the system

$$\frac{\partial \psi_j(s, t)}{\partial t} = \sum \tilde{P}_{jk}(s, t)\psi_k(s, t)$$

is of order one and of type not larger than $b(t - t_0)$. Replacing the variable s by the operator L, one obtains (on the basis of the preceding theorem) an operator defined on $LS_{\alpha,A}^{1,B}$, for sufficiently small $t - t_0$, which maps this space into $LS_{\alpha,A}^{1,Be}$. At the same time, $Q(L, t_0, t)$ is the resolvent operator for the Cauchy problem of the inverse-adjoint of the systems (1) and (2). This can be proved by the methods known from Section 3.

Thus, the inverse-adjoint of the Cauchy problem (1), (2) has a resolvent mapping the test function space $LS_{\alpha,A}^{1,B}$ into $LS_{\alpha,A}^{1,Be}$.

Let E denote the space consisting of measurable functions $\varphi(x)$ endowed with the norm

$$\|\varphi\|_E = \int_{-\infty}^{\infty} \exp[a\,|x|^{p_1'}]\,\|\varphi(x)\|\,dx.$$

For $a < A$, $LS_{\alpha,A}^{1,B}$ (as well as $LS_{\alpha,A}^{1,Be}$) is part of the space E, and it follows from the above that it is a *dense* subset of E.

We are thus within the requirements of the theorem of Section 2.2: We have three spaces $\Phi \subset \Phi_1 \subset E$ with Φ everywhere dense in E, the operator $Q(L, t_0, t)$ maps Φ into Φ_1 and the inverse-adjoint of the Cauchy problem (1), (2) has a solution belonging to Φ_1 for arbitrary initial functions $\varphi_j(x, t) \in \Phi$.

Specialized to the case under consideration the theorem yields the following result: The Cauchy problem for the system admits at most one solution, belonging to the space E' for all t $(0 \leqslant t \leqslant T)$. In particular the system admits at most one solution with

$$| u_k(x, t) | \leqslant C_1 \exp[b \, | x |^{p_1'}].$$

The operator

$$Lu = \Delta u + q(x)u,$$

where Δ is the Laplacian, and $q(x)$ is a continuous function which is bounded everywhere together with its first order partial derivatives, is an example of an elliptic operator. Here the number p_1 is 2. Our results imply that for any equation of the form

$$\frac{\partial u}{\partial t} = \alpha \, \Delta u + q(x)u,$$

with α an arbitrary (complex) constant, the functions $f(x)$ which satisfy an inequality

$$| f(x) | \leqslant C \exp[b \, | x |^2],$$

form a uniqueness class for the corresponding Cauchy problem.

CHAPTER III

CORRECTNESS CLASSES FOR THE CAUCHY PROBLEM

1. Introduction

We consider the system of partial differential equations

$$\frac{\partial u_j(x, t)}{\partial t} = \sum_{k=1}^{m} P_{jk}\left(i \frac{\partial}{\partial x}\right) u_k(x, t) \qquad (j = 1, 2, \ldots, m), \tag{1}$$

with the initial (Cauchy) conditions

$$u_j(x, 0) = u_j(x). \tag{2}$$

In Chapter II we have found the uniqueness classes for the Cauchy problem for all such systems. We have left open the problem of existence of solutions.

In this chapter we shall indicate for each system of the form (1) a natural class of initial conditions (Cauchy data), guaranteeing the existence of a solution for the system (1) and (2), which depends continuously on the Cauchy data. Since in each case the considerations will be carried out within the uniqueness classes appropriate for the given system, as established in Chapter II, the solution we thus find is also unique.

We shall make use in this chapter of the method of Fourier transforms for generalized functions.

The system (1) should be considered as a system in unknown *generalized functions* $u_j(x, t)$, i.e., linear continuous functionals over a test function space Φ (corresponding to the appropriate uniqueness class for the Cauchy problem). After subjecting the system (1) to a Fourier transformation in the variables x, we obtain the system of ordinary differential equations

$$\frac{\partial v_j(s, t)}{\partial t} = \sum P_{jk}(s) v_k(s, t) \tag{3}$$

with the initial condition

$$v_j(s, 0) = v_j(s) = \widetilde{u_j(x)}; \tag{4}$$

here $v_j(s, t)$ is to be considered a linear functional over the space $\Psi = \tilde{\Phi}$. It will be shown (cf. Theorem 1, infra) that *the solution of this Cauchy problem* (which is unique in the given class) *is, in vector notation,*

$$v(s, t) = Q(s, t) \cdot v(s, 0),$$

where (5)

$$Q(s, t) = e^{tP(s)}$$

is the resolvent matrix-function for the system (3).

Making use of the theorem on the Fourier transform of a product (Section 3, Chapter III, Volume 2) the inverse Fourier transform of the solution (5) yields

$$u(x, t) = G(x, t) * u(x, 0), \tag{6}$$

where $G(x, t)$ is the *Green's matrix*[1] (or matrix Green's function), i.e. the inverse Fourier transform of the matrix-function $Q(s, t)$; the kth column of the Green's matrix yields a solution of the Cauchy problem (1) and (2) with Cauchy data of the form $u_j(x, 0) = 0$ for $j \neq k$, $u_k(x, 0) = \delta(x)$.

In general, Eq. (6) defines a generalized function. It is one of the purposes of this chapter to clarify under what conditions the solution of the Cauchy problem is an ordinary function.

For this purpose we shall subject the matrix-function $G(x, t)$ to a detailed analysis and shall impose appropriate restrictions on the initial function $u(x, 0)$. The better "behaved" the function $G(x, t)$ is, the weaker will be the conditions to be imposed on the initial functions $u(x, 0)$. It turns out that if the matrix $G(x, t)$ is made up of ordinary functions, we do not have to impose any smoothness conditions on $u(x, 0)$, and it is sufficient to require that the growth of these functions as $|x| \to \infty$ should not be too rapid in order to guarantee the convergence of the integral which defines the convolution (6). If the Green's matrix $G(x, t)$ is made up of generalized functions of order h (i.e., which are derivatives of order h of an ordinary function) then the existence of the convolution (6) as an ordinary function is guaranteed by requiring that the initial functions $u(x, 0)$ admit derivatives up to order h. Finally, if the matrix

[1] The Green's matrix is sometimes called the "elementary solution" of the system, the term Green's function or matrix being reserved for the nonhomogeneous equation, with a delta-function in the right-hand side. (Translator's Note.)

$G(x, t)$ consists of generalized functions of "infinite order" (i.e., generalized functions which are not derivatives of finite order of ordinary functions), one has to require that the initial functions be infinitely often differentiable, or even analytic. The indicated possibilities are of course determined by the structure of the resolvent matrix $Q(s, t)$, in particular by its behavior in the complex plane of the parameter s. In turn, as we have seen in Chapter II, the behavior of the matrix $Q(s, t)$ depends essentially on the behavior of the characteristic roots $\lambda_j(s)$ of the matrix $P(s)$. Accordingly, we distinguish three classes of systems:

1. Parabolic Systems

These systems are characterized by the fact that the function $\Lambda(s) = \max_j \operatorname{Re} \lambda_j(s)$ satisfies for real $s = \sigma$ the inequality

$$\Lambda(\sigma) < -C \mid \sigma \mid^h + C_1$$

with $C > 0$, $h > 0$. It turns out that in this case, the elements of the matrix $G(x, t)$ are ordinary functions which decrease exponentially for $| x | \to \infty$. Therefore, the solution of a parabolic system exists for any (locally integrable) initial data $u(x, 0)$, with no restriction whatsoever on their smoothness (differentiability) and which may even have a corresponding exponential increase for $| x | \to \infty$. Quantitatively one can express this increase in terms of the reduced order p_0 of the system, the parabolicity exponent h and a characteristic of the function $\Lambda(s)$ which will be called the *genus* of the system and will be discussed below in Section 2. The convolution (6) then becomes an analog of the Poisson formula for the heat equation.

2. Systems Which Are Correct According to Petrovskiĭ (Petrovskiĭ-Correct Systems)

These systems are characterized by the fact that the function $\Lambda(s) = \max_j \operatorname{Re} \lambda_j(s)$ is bounded for real $s = \sigma$:

$$\Lambda(\sigma) \leqslant C.$$

In this case the matrix elements of $G(x, t)$ are generalized functions of finite fixed order l (i.e., derivatives of fixed order l of ordinary functions). Therefore the solution of such a correct system exists for initial functions admitting derivatives up to an order determined by the number l. The growth properties of these functions for $| x | \to \infty$ depend on the system (1); the initial functions can be either functions of arbitrary growth (hyperbolic systems) or functions with exponential growth (regular systems) or, finally, functions of only power law growth (systems of the Schrödinger type). This behavior is also determined by the function $\Lambda(s)$.

3. INCORRECT SYSTEMS

These systems are characterized by the fact that the function $\Lambda(s)$ actually increases according to a power law along the real axis. The convolution (6) will be an ordinary function only if the initial functions are infinitely differentiable or analytic. If the inequality

$$\Lambda(\sigma) \leqslant C \mid \sigma \mid^{h} + C_1$$

holds for $h < 1$, the initial functions must be infinitely differentiable with some additional conditions on the growth of their derivatives. In the general case, when

$$\Lambda(\sigma) \leqslant C \mid \sigma \mid^{p_0} + C_1 ,$$

the initial data must be analytic, of a definite order as $| z | = | x + iy | \to \infty$.

Let us return to the formulation of our fundamental theorem. We consider the system of partial differential equations (in vector notation)

$$\frac{\partial u(x, t)}{\partial t} = P\left(i \frac{\partial}{\partial x}\right) u(x, t) \tag{7}$$

with the initial condition

$$u(x, 0) = u_0(x). \tag{8}$$

The unknowns $u(x, t)$ are considered generalized functions over some test function space Φ, chosen in such a manner that the solution of the Cauchy problem should exist and be unique in the dual space Φ' (the space of generalized functions). We have indicated in Chapter II, Sections 3 and 4 how to construct such spaces. After performing a Fourier transformation we obtain the equivalent problem

$$\frac{dv(s, t)}{dt} = P(s)v(s, t), \tag{9}$$

$$v(s, 0) = v_0(s), \tag{10}$$

where $v(s, t)$ is a generalized function defined on the Fourier-dual $\Psi = \tilde{\Phi}$.

Theorem 1. *The Cauchy problem* (9) *and* (10) *admits the solution* (still in vector notation)

$$v(s, t) = Q(s, t)v(s, 0), \tag{11}$$

where

$$Q(s, t) = e^{tP(s)}$$

is the resolvent matrix-function of the system (9).

Proof. As was shown in Chapter II, the adjoint operator of multiplication by the function

$$\exp[tP^*(s)]$$

is defined and bounded in the space Ψ (for $0 \leqslant t \leqslant T$) and maps that space into a wider space Ψ_1. The same is true also for the operator

$$P^*(s) \exp[tP^*(s)]$$

which is obtained by formal differentiation with respect to t of the first operator. Therefore both operators

$$e^{tP(s)} \quad \text{and} \quad P(s)\, e^{tP(s)}$$

are defined and bounded in the space Ψ_1' and map this space into Ψ'.

For $\Delta t \to 0$ the operator

$$\frac{\Delta \exp[tP^*(s)]}{\Delta t} = \frac{\exp[(t + \Delta t)P^*(s)] - \exp[tP^*(s)]}{\Delta t}$$

converges (for each test function $\psi(s)$) to the operator $P^*(s) \exp[tP^*(s)]$, since the functions

$$\frac{\exp[(t + \Delta t)P^*(s)] - \exp[tP^*(s)]}{\Delta t}\, \psi(s) = P^*(s) \exp[(t + \theta \Delta t)P^*(s)]\, \psi(s)$$

are bounded in the topology of Ψ_1 and converge regularly to $P^*(s) \exp[tP^*(s)]\, \psi(s)$. Therefore the adjoint operator in the space Φ_1' satisfies the relation

$$\frac{\exp[(t + \Delta t)P(s)] - \exp[tP(s)]}{\Delta t} \to P(s) \exp[tP(s)]$$

for each generalized function $v_0(s)$. This implies that the expression

$$v(s, t) = \exp[tP(s)]\, v_0(s)$$

defines a solution of the equation (11). For $t \to 0$, the operator $\exp[tP^*(s)]$ converges to the unit operator on each test function $\psi(s)$ (due to the boundedness in Ψ and the regular convergence of the functions $\exp[tP^*(s)]\,\psi(s)$ for $t \to 0$). It follows that the adjoint operator $e^{tP(s)}$ converges to the unit operator on each generalized function $v_0(s)$. In other words, in the topology of Ψ'

$$e^{tP(s)}v_0(s) \to v_0(s).$$

Thus the solution (11) satisfies the initial condition.

It should be noted that *the elements of the matrix* $Q(s, t)$ *are multiplication operators (multipliers), mapping the space* $\Psi' = \tilde{\Phi}'$ *into itself.*

In agreement with the preceding, the generalized function

$$u(x, t) = G(x, t) * u_0(x)$$

is a solution of the system (7), satisfying the initial condition (8). This solution depends continuously on the Cauchy datum $u_0(x)$: if the initial function $u_0(x)$ depends on a parameter λ and u_0 converges to zero for $\lambda \to 0$, as a functional on the test function space Φ, the solution $u(x, t)$, which also depends on the parameter λ, converges to zero in the topology of the dual space of Φ, for every fixed $t > 0$.

The proof is carried out by means of Fourier transformation. The solution of the Fourier-transformed system

$$v(s, t) = e^{tP(s)}v_0(s)$$

also depends on the parameter λ. Due to the continuity of the Fourier transform, $u_0(x) \to 0$ implies $v_0(s) \to 0$ (in the space Ψ'). It further follows from the properties of the multiplier $e^{tP(s)}$ that for $\lambda \to 0$ we also have $e^{tP(s)}v_0(s) = v(s, t) \to 0$ in Ψ'. Carrying out the inverse Fourier transformation, we find that $u(x, t) \to 0$ in Φ', as required.

The following important circumstance requires attention. Had we shown that the solution of the system (1), written in the form (6)

$$u(x, t) = G(x, t) * u(x, 0)$$

is an ordinary function, this would not imply that $u(x, t)$ is also a solution of the system (1) in the usual sense, since $u(x, t)$ might not admit derivatives in the conventional sense, either with respect to x or with respect to t. This function is a *solution of the system* (1) *in the generalized function (distribution) sense*: i.e., for each test function $\varphi(x)$

$$\frac{\partial}{\partial t}\int u(x, t)\varphi(x)\,dx = \int u(x, t)P^*\left(i\frac{\partial}{\partial x}\right)\varphi(x)\,dx. \tag{12}$$

If the function $u(x, t)$ is sufficiently smooth, so that the operator $\partial/\partial t$ in the left-hand side of (12) can be taken under the $\int$ sign, and the right-hand side can be subjected to an integration by parts, transposing the differential operator onto the function $u(x, t)$, only then is the function $u(x, t)$ also a solution in the conventional sense for the system (1).

Sufficient smoothness (differentiability) of the function $u(x, t)$ in the variable x can be ensured by imposing additional smoothness requirements on the initial function $u(x, 0)$. As regards the smoothness of the function $u(x, t)$ in the variable t, it is completely determined by the properties of the function $G(x, t)$, and there does not seem to exist a general method for investigating it.

The same is true for the verification of the initial condition. The solution $u(x, t)$ which has been constructed verifies this condition $u(x, 0) = \lim_{t\to 0} u(x, t)$ in the sense of generalized functions (i.e., the limit is taken in the space Φ'); this means that for every test function $\varphi(x)$ we have

$$\lim_{t\to\infty} \int u(x, t)\varphi(x)\, dx = \int u(x, 0)\varphi(x)\, dx. \tag{13}$$

It seems likely that the equality (13) is in itself the most correct expression of the fact that the initial condition is satisfied.

2. Parabolic Systems

2.1. Definition and Examples

A system of order p with constant coefficients

$$\frac{\partial u_j(x, t)}{\partial t} = \sum_{k=1}^{m} P_{jk}\left(i\frac{\partial}{\partial x}\right) u_k(x, t) \tag{1}$$

is said to be *parabolic*, if the function $\Lambda(s) = \max_j \operatorname{Re} \lambda_j(s)$ satisfies for real $s = \sigma$ the inequality

$$\Lambda(\sigma) \leqslant -C\,|\,\sigma\,|^h + C_1, \qquad C > 0, \qquad h > 0. \tag{2}$$

The number h is called the *parabolicity exponent* (index) of the system (1).

Example 1. The heat equation

$$\frac{\partial u}{\partial t} = \frac{\partial^2 u}{\partial x^2};$$

here

$$\lambda(s) = -s^2, \qquad \Lambda(\sigma) = \lambda(\sigma) = -\sigma^2, \qquad C = 1, \quad h = 2.$$

Example 2. The equation

$$\frac{\partial u}{\partial t} = \frac{\partial^2 u}{\partial x^2} + i\left(i\frac{\partial}{\partial x}\right)^p u, \qquad p > 2; \tag{3}$$

with

$$\lambda(s) = -s^2 + is^p, \qquad \Lambda(\sigma) = \operatorname{Re}\lambda(\sigma) = -\sigma^2, \quad h = 2.$$

Example 3. *Systems which are parabolic in the sense of Petrovskiĭ* (Petrovskiĭ-parabolic). We decompose the matrix $P(s)$ of the system (1) into a sum of $\hat{P}(s)$ and $R(s)$, such that the first matrix contains only the highest order terms (terms of order p) and the second contains all the lower order terms.

Let $\hat{\lambda}_1(s),\dots,\hat{\lambda}_m(s)$. denote the characteristic roots of the matrix $\hat{P}(s)$. A system (1) is said to be *Petrovskiĭ-parabolic* if for $|\sigma| = 1$ the quantities $\operatorname{Re}\hat{\lambda}_j(\sigma)$ are bounded by a fixed negative constant $-\omega$. We show that *each system which is Petrovskiĭ-parabolic is also parabolic in our sense and the parabolicity index h for such systems coincides both with the order p and with the reduced order p_0 of the system.*

Proof. Let the characteristic roots $\hat{\lambda}_1(s),\dots,\hat{\lambda}_m(s)$ of the matrix $\hat{P}(s)$ satisfy for $|\sigma| = 1$ the inequality

$$\operatorname{Re}\lambda_j(\sigma) < -\omega.$$

We set $\xi = \sigma/|\sigma|$, $\gamma_j(\sigma) = \hat{\lambda}_j(\sigma)/|\sigma|^p$. Factoring out $|\sigma|^p$ from all elements of the determinant $\det\|P(\sigma) - \lambda E\|$ and dividing by this factor, we obtain an equation of the form

$$\det\|\hat{P}(\xi) + \epsilon(\sigma) - \gamma E\| = 0, \qquad \gamma = \lambda/|\sigma|^p,$$

where $\epsilon(\sigma) \to 0$ as $|\sigma| \to \infty$. The zeros of this equation converge for $|\sigma| \to \infty$ and fixed $\sigma/|\sigma| = \xi$ to the zeros of the equation

$$\det\|\hat{P}(\xi) - \gamma E\| = 0$$

uniformly with respect to ξ for $|\xi| = 1$ (otherwise there would be a contradiction with the theorem on continuous dependence of the roots of an equation, where the coefficient of the highest power is one, on the other coefficients). Thus, for sufficiently large $|\sigma| \geqslant \sigma_0$

$$\left|\frac{\lambda_j(\sigma)}{|\sigma|^p} - \hat{\lambda}_j(\xi)\right| < \frac{\omega}{2},$$

and since $\operatorname{Re} \hat{\lambda}_j(\xi) < -\omega$, we have

$$\operatorname{Re} \lambda_j(\sigma) < |\sigma|^p \operatorname{Re} \hat{\lambda}_j(\xi) + \tfrac{1}{2}\omega |\sigma|^p < -\tfrac{1}{2}\omega |\sigma|^p.$$

Therefore for arbitrary σ we have the estimate

$$\Lambda(\sigma) = \max \operatorname{Re} \lambda_j(\sigma) < -\tfrac{1}{2}\omega |\sigma|^p + C.$$

Consequently, the system (1) is indeed parabolic with parabolicity index p.

As a consequence of the note at the end of Section 6.1, Chapter II, the function $\Lambda(s)$ has for complex s a (power-law) degree of growth not lower than the order of its decrease for real $s = \sigma$. The power-law order of the function $\Lambda(s)$ for complex s is the reduced order p_0, hence $p \leqslant p_0$. On the other hand, it was shown in Section 3, Chapter II, that always $p \leqslant p_0$. Hence, in the case under consideration, we have

$$h = p = p_0,$$

as asserted.

We further note that the heat equation (Example 1) is parabolic according to Petròvskiĭ, whereas the parabolic equation (3) does not belong to the class of equations which are parabolic in the Petrovskiĭ sense.

2.2. The Resolvent Matrix

According to the general method described in Section 1, we shall solve the Cauchy problem for the system (1) of Section 2.1 by means of Fourier transforms. The transformed system becomes a system of ordinary differential equations

$$\frac{dv(s, t)}{dt} = P(s)v(s, t), \tag{1}$$

with the resolvent matrix

$$Q(s, t) = e^{tP(s)}.$$

We estimate the resolvent matrix $Q(s, t)$ for the case of a parabolic system.

First of all we have the estimate (cf. Section 3.2, Chapter II)

$$\| Q(s, t)\| \leqslant C(1 + |s|)^{p(m-1)} \exp[bt |s|^{p_0}], \tag{2}$$

since the system has reduced order p_0.

Making use of the parabolicity condition and the inequality (6) in Section 6.1, Chapter II, we obtain for real $s = \sigma$

$$\| Q(\sigma, t)\| \leqslant C_1(1 + |\sigma|)^{p(m-1)} \exp[-at |\sigma|^h]. \tag{3}$$

Remark. Conversely, if the inequality (3) holds, the same fundamental relation (6) in Section 6.1, Chapter II implies that

$$\Lambda(\sigma) \leqslant - C |\sigma|^h + C_1 ,$$

i.e., that the system under consideration is parabolic. Thus one could use the inequality (3) as a definition of parabolicity of a system.

All the following constructions will be based only on the fact that the inequalities (2) and (3) hold. This allows one to include also systems with variable coefficients (depending on t) and also more general systems (e.g., convolution equations) which upon Fourier transformation reduce to the form:

$$\frac{dv(s, t)}{dt} = P(s, t)v(s, t)$$

with a resolvent matrix $Q(s, t)$ satisfying the inequalities (2) and (3). In the sequel we shall call all such systems *parabolic systems.*

2.3. The Genus of a Parabolic System

Lemma 1. *Every parabolic system has a reduced order p_0 strictly larger than* 1.

Proof. We consider the sum of all characteristic roots $\lambda_1(s)+\cdots+\lambda_m(s)$ of the matrix $P(s)$. Being one of the coefficients of the characteristic polynomial $\det \| P(s) - \lambda E \|$, it is itself a polynomial in $s_1 ,..., s_n$. Its real part $R(s_1 ,..., s_n)$ is a polynomial in $\sigma_1 , \tau_1 ,..., \sigma_n , \tau_n$. Assuming that $p_0 \leqslant 1$, this means that the polynomial $R(s_1 ,..., s_n)$ increases for $|s| \to \infty$ not faster than $|s|$, and therefore is at most of first degree. For real $s = \sigma$, it is consequently a real function of the arguments $\sigma_1 ,..., \sigma_n$:

$$R(\sigma_1 ,..., \sigma_n) = a_0 + \sum_{k=1}^{n} a_k \sigma_k .$$

But a linear function cannot go to $-\infty$ when $|\sigma| \to \infty$. Therefore, the system under consideration cannot be parabolic, proving our assertion.

Let us apply the Theorem 1 in Section 7, Chapter IV, Volume 2 to the functions $Q(s, t)$ for fixed t. It follows from that theorem that there exists a domain H_μ determined by the inequality

$$|\tau| \leqslant K(1 + |\sigma|)^\mu, \qquad \mu \geqslant 1 - (p_0 - h), \tag{1}$$

where

$$\|Q(\sigma + i\tau, t)\| \leqslant C \exp[-a't\,|\sigma|^h], \tag{2}$$

and the constant a' differs by arbitrarily little from a.[2]

The least upper bound of the numbers μ satisfying the inequalities (1) and (2) is one of the most important characteristics of a parabolic system. We shall call this number the *genus* of the system.

We do not know whether the l.u.b. is attained in the class of all admissible numbers μ. The subsequent derivations will be carried out, for simplicity, under the assumption that this l.u.b. is attained, so that μ in (1) can be considered equal to the genus of the system. Should this not be true in reality, one could choose as μ in the subsequent constructions an arbitrary number which is smaller than the genus of the system, with appropriate modifications of the formulations of the statements.

We shall see below that the *correctness class for the Cauchy problem of a parabolic system is determined by the genus of the system.* This class consists of those functions $f(x)$ which for $|x| \to \infty$ have an exponential growth of order $\leqslant p_1 = p_0/(p_0 - \mu)$ for systems of positive genus ($\mu > 0$) or of order $\leqslant p_2 = h/(h - \mu)$ for systems of nonpositive genus ($\mu \leqslant 0$). In the first case, the admissible order is a number larger than one (for systems of positive genus) and in the second case (nonpositive genus) this order is smaller than or equal to one.

One can gain information about the genus of a parabolic system with constant coefficients by considering its characteristic roots. On the basis of the fundamental inequality (6) of Section 6.1, Chapter II, which connects the growth of the function $Q(s, t)$ with the growth of the function $\Lambda(s) = \max_j \operatorname{Re} \lambda_j(s)$ the genus of a parabolic system may also be defined as the largest exponent, such that in the region

$$|\tau| \leqslant K(1 + |\sigma|)^\mu$$

the function $\Lambda(s)$ satisfies the inequality

$$\Lambda(s) \leqslant -C|\sigma|^h + C_1.$$

[2] The parameter t does not occur in the formulation of the indicated theorem. But the constant a' can be taken in the form $a(1 - \epsilon)$ for arbitrary $\epsilon > 0$, and it is easy to see that in the presence of t one can replace the coefficient at by $at(1 - \epsilon) = a't$: the region H_μ does not depend on t, as is clear from its definition.

Note. It was already stated that

$$\mu \geqslant 1 - (p_0 - h).$$

The number μ can in reality exceed $1 - (p_0 - h)$.

Example. Assume that the characteristic roots of the matrix $P(s, t)$ are of the form

$$\lambda_1(s) = is^6 - s^4, \qquad \lambda_2(s) = is^4 - s^2. \tag{3}$$

In this case $p_0 = 6$, $h = 2$ and Theorem 1, Section 7, Chapter IV, Volume 2 yields $\mu \geqslant -3$. In fact the function $\Lambda(s) = \max \operatorname{Re}\{\lambda_1(s)\ \lambda_2(s)\}$ decreases according to a power law in the region $|\tau| \leqslant K(1 + |\sigma|)^{-1}$ (with the exponent -1, instead of -3). In order to see this we expand the real parts of the roots λ_1 and λ_2

$$\operatorname{Re}\lambda_1(s) = -6\tau\sigma^5 - \sigma^4 + \ldots, \qquad \operatorname{Re}\lambda_2(s) = -4\tau\sigma^3 - \sigma^2 + \ldots,$$

where the dots denote lower powers of σ.

In the region $|\tau| \leqslant K(1 + |\sigma|)^{-1}$ with sufficiently large $K > 0$ the first term cannot be larger than $-C_1\sigma^4$ and the second term cannot be larger than $-C_2\sigma^2$. This implies that in the indicated region $\Lambda(s)$ does not exceed $-C_2\sigma^2 + C_3$. Therefore one may take for μ the value -1.

V. M. Borok has shown that for systems with one space variable one can construct simple formulas for the computation of all characteristics of the system. Every zero of the equation $\det \| P(s) - \lambda E \| = 0$ can be expanded in the neighborhood of the point at infinity into the series[3]

$$\lambda(s) = \alpha_0 s^{k_0} + \alpha_1 s^{k_1} + \cdots + \alpha_p s^{k_p} + \ldots,$$

$(\alpha \neq 0)$ with $k_0 > k_1 > \cdots$ rational exponents (which can be determined by means of the Newton polygon). Among the coefficients α_q there are certainly some for which $\operatorname{Re}\alpha_q \neq 0$. Let $\operatorname{Re}\alpha_0 = \operatorname{Re}\alpha_1 = \cdots = \operatorname{Re}\alpha_{p-1} = 0$, but $\operatorname{Re}\alpha_p \neq 0$. It can be shown that $k_0, k_1, \ldots, k_{p-1}$ are positive integers, and k_p is a positive even integer. The following formulas are true

$$p_0 = \max k_0 \qquad h = \min k_p \qquad \mu = \min\{k_p - k_0 + 1\}$$

(with respect to all roots)

[3] Cf. N. G. Chebotarev, "Teoriya algebraicheskikh funktsii" (Theory of Algebraic Functions), Chapter IV. Gostekhizdat, Moscow-Leningrad, 1948.

In particular, the following three characteristic quantities: the reduced order p_0 the parabolicity exponent h and the genus μ are integers in the case of one space variable (in addition h is an even integer).

2.4. The Fundamental Theorem for Systems with Positive Genus

Theorem 1. *Consider the parabolic system*

$$\frac{\partial u_j(x, t)}{\partial t} = \sum_{k=1}^{m} P_{jk}\left(i\frac{\partial}{\partial x}\right) u_k(x, t) \qquad (j = 1, \ldots, m) \tag{1}$$

of positive genus $\mu > 0$. *If the initial functions* $\{u_j(x, 0)\} = u_0(x)$ *of this system belong to the class* K_{p_1, b_0} *of functions* $f(x)$ *which satisfy the inequality*

$$|f(x)| \leqslant C \exp[b_0 |x|^{p_1}], \quad \text{with} \quad p_1 = \frac{p_0}{p_0 - \mu}, \tag{2}$$

then, for sufficiently small $t > 0$ *and arbitrarily given* $b_1 > b_0$, *the solution of the system belongs to the class* K_{p_1, b_1}.

Proof. We first assume that the number of independent variables is $n = 1$. Since $p_1 = p_0/(p_0 - \mu) \leqslant p_0(p_0 - 1)$,[4] the class K_{p_1, b_1} is contained, for sufficiently small T, in the uniqueness class of the Cauchy problem for the system (1). Within the uniqueness class the solution of the Cauchy problem for the system (1) can be written as a convolution

$$u(x, t) = G(x, t) * u(x, 0), \tag{3}$$

with $G(x, t)$ the Green's matrix—i.e., the inverse Fourier transform of the resolvent matrix of the system of ordinary differential equations

$$\frac{dv_j(s, t)}{dt} = \sum_{k=1}^{m} P_{jk}(s) v_k(s, t) \qquad (j = 1, \ldots, m). \tag{4}$$

In general, the convolution (3) transforms ordinary functions $u(x, 0)$ into generalized functions $u(x, t)$. We have to show that the convolution maps the class K_{p_1, b_0} into the class K_{p_1, b_1}, i.e., that the ordinary functions belonging to the class K_{p_1, b_0} are taken by the convolution into *ordinary functions* belonging to the class K_{p_1, b_1}.

For this purpose we construct the matrix function $G(x, t)$. Its Fourier transform coincides with the resolvent matrix $Q(s, t)$ of the system (4), hence we shall investigate the properties of this latter matrix. For $Q(s, t)$

[4] The inequality $\mu \leqslant 1$ has in fact been proved in Section 7, Chapter IV, Volume 2.

we have derived in Section 2.2 the estimates (2) and (3), and in addition we know that $\mu > 0$, by assumption. According to Theorem 2 in Section 7, Chapter IV, Volume 2, for $t > 0$ the matrix $Q(s, t)$ satisfies the inequality

$$\| Q(s, t) \| \leqslant C' \exp[-at \mid \sigma \mid^h + b't \mid \tau \mid^{p_0/\mu}], \tag{5}$$

where the constant b' is not larger than $B_1(a + b_0)$. B_1 depends only on the region H_μ, but not on t.

Assume that the time coordinate t varies between 0 and T. Then the inequality (5) implies that the matrix elements of $Q(s, t)$ belong to the space $W^{p_0/\mu, \theta}$ consisting of all entire functions $\varphi(s)$ which admit the majorizations

$$| s^k \varphi(\sigma + i\tau) | \leqslant C_k \exp\left[\frac{\mu}{p_0} \mid \bar{\theta}_\tau \mid^{p_0/\mu}\right],$$

for arbitrary $\bar{\theta} > \theta$ and $(\mu/p_0)\, \theta^{p_0/\mu} = b'T$ (cf. Section 3, Chapter II).[5] According to the theorem on Fourier transforms of the space $W^{p_0/\mu, \theta}$ (Section 3, Chapter I) the elements of the matrix $Q(x, t)$ belong to the Fourier dual space

$$\widetilde{W^{p_0/\mu, \theta}} = W_{(p_0/\mu)', 1/\theta},$$

with $(p_0/\mu)'$ defined by the equation

$$\frac{1}{(p_0/\mu)} + \frac{1}{(p_0/\mu)'} = 1,$$

Consequently

$$(p_0/\mu)' = p_0/(p_0 - \mu) = p_1 .$$

It follows from Section 1, Chapter I that the functions $\varphi(x)$ belonging to a space $W_{p,a}$ satisfy the inequality

$$| \varphi(x) | \leqslant C \exp\left[-\frac{1}{p} \mid \bar{a}x \mid^p\right], \qquad \bar{a} < a \qquad \text{arbitrary.}$$

Thus, in the case under consideration

$$\| G(x, t) \| \leqslant C \exp\left(-\frac{1}{p_1}\left|\frac{x}{\bar{\theta}}\right|^{p_1}\right), \qquad \bar{\theta} > \theta \qquad \text{arbitrary.} \tag{6}$$

[5] In reality the elements of the matrix $Q(s, t)$ belong to the smaller space $W^{p_0/\mu, \theta}_{h, \theta'}$, but we cannot fix θ' independent of t in the interval $(0, T)$. The notations for the spaces used here agree with those of the end of Section 3, Chapter I.

It should be explicitly understood that this estimate is valid for all t in the closed interval $0 \leqslant t \leqslant T$ where $(\mu\, p_0)\, \theta^{p_0/\mu} = b'T$.

Thus $G(x, t)$ is an ordinary function, with exponential decrease. We consider now the ordinary convolution

$$G(x, t) * u(x, 0) = \int G(\xi - x, t) u_0(\xi)\, d\xi$$
$$= \int G(\xi, t) u_0(x - \xi)\, d\xi, \tag{7}$$

and show that for sufficiently small T this convolution exists and belongs to the class K_{p_1, b_1}, with $p_1 = p_0(p_0 - \mu)$ and $b_1 > b_0$. We shall prove then that the ordinary convolution coincides with the convolution (3) in the sense of generalized functions.

We first establish a lemma.

Lemma 2. *Let* $\lambda > 0$. *For arbitrary* $\gamma > \beta > 0$ *there exists a number* $\alpha > 0$, *such that for all* x *and* ξ *there is an inequality*

$$-\alpha \mid \xi \mid^{\lambda} + \beta \mid x - \xi \mid^{\lambda} \leqslant \gamma \mid x \mid^{\lambda}. \tag{8}$$

Proof. For $\rho = x/\xi$ the inequality (8) is transformed into the equivalent inequality

$$\gamma \mid \rho \mid^{\lambda} - \beta \mid 1 - \rho \mid^{\lambda} \geqslant -\alpha. \tag{9}$$

For $\gamma > \beta$, $\gamma \mid \rho \mid^{\lambda} - \beta \mid 1 - \rho \mid^{\lambda}$ is a continuous function of ρ which increases without bound for $\mid \rho \mid \to \infty$. Consequently the function is bounded from below so that for $\alpha > 0$ the inquality (9) is verified. But then (8) is also true.

We now go over to the proof of Theorem 1. The integrand in

$$\int G(\xi, t) u_0(x - \xi)\, d\xi \tag{10}$$

admits the majorant

$$\exp\left[-\frac{1}{p_1} \mid \bar{\theta}^{-1} \xi \mid^{p_1} + b_0 \mid x - \xi \mid^{p_1}\right]$$

(this follows from (9) and the assumption made about $u_0(x - \xi)$). According to the lemma $(1/2p_1)\bar{\theta}^{-p_1}$ can be chosen so large that either

$$-\frac{1}{2p_1} \bar{\theta}^{-p_1} \mid \xi \mid^{p_1} + b_0 \mid x - \xi \mid^{p_1} \leqslant b_1 \mid x \mid^{p_1},$$

or

$$\exp\left[-\frac{1}{p_1}\bar{\theta}^{-p_1}|\xi|^{p_1}+b_0|x-\xi|^{p_1}\right]\leqslant\exp\left[-\frac{1}{2p_1}\bar{\theta}^{-p_1}|\xi|^{p_1}\right]\cdot\exp[b_1|x|^{p_1}].$$

be true. Consequently the integral (10) converges if θ is selected as indicated, and the integral satisfies the inequality

$$\left|\int G(\xi,t)u_0(x-\xi)\,d\xi\right|\leqslant\exp[b_1|x|^{p_1}]\int\exp\left[-\frac{1}{2p_1}\bar{\theta}^{-p_1}|\xi|^{p_1}\right]d\xi$$
$$=C\exp[b_1|x|^{p_1}]. \tag{11}$$

The choice of $(1/2p_1)\,\bar{\theta}^{-p_1}$ fixes also the interval $0\leqslant t\leqslant T$, since $(\mu/p_0)\,\theta^{p_0/\mu}=b'T$ and $\bar{\theta}>\theta$. For b_1 converging to b_0 the number $\bar{\theta}^{-p_1}$ increases without bound, and the interval $0\leqslant t\leqslant T$ contracts to zero.

We have thus shown that the convolution (7) exists in the ordinary sense for sufficiently small T. We now show that the convolution (3), considered as a convolution in the sense of the theory of generalized functions, coincides with (7).

Applying the convolution (3) to the test function $\varphi(x)$ yields

$$(u,\varphi)=(G*u_0,\varphi)=(u_0,G*\varphi)=\left(u_0(x),\int G(\xi,t)\varphi(x+\xi)\,d\xi\right)$$
$$=\int u_0(x)\left\{\int G(\xi,t)\varphi(x+\xi)\,d\xi\right\}dx. \tag{12}$$

The convolution (7) acts on these same test functions $\varphi(x)$ as follows

$$(G(x,t)*u_0(x),\varphi(x))=\int\left\{\int G(x,t)u_0(\eta-x)\,dx\right\}\varphi(\eta)\,d\eta. \tag{13}$$

Replacing ξ by $\eta-x$, the integral (22) goes over into the integral (13), if one changes the order of integrations, which is allowed on the basis of Fubini's theorem, since the double integral

$$\iint|G(x,t)u_0(\eta-x)\varphi(\eta)|\,dx\,d\eta$$

converges, by the above estimates.

Consequently the solution

$$u(x,t)=G(x,t)*u_0(x)=\int G(\xi-x)u_0(\xi)\,d\xi$$

is an ordinary function, belonging to the class K_{p_1,b_1} with arbitrarily small $b_1 - b_0$, as asserted.

The proof can be adapted to the case of n space variables $x_1, \ldots, x_n$, by making the following modifications:

Wherever theorems from Section 7, Chapter IV, Volume 2 have been used, one has to use their n-dimensional analogs (Cf. Section 9, Chapter IV, Volume 2).

In the formulation of the lemma x and ξ have to be replaced by n-dimensional vectors. The inequality to be proved has to be divided by $|\xi|$ rather than by ξ. Then the inequality

$$\gamma |\rho|^\lambda + \beta |e - \rho|^\lambda \geqslant -\alpha,$$

here ρ is an arbitrary vector, e is a unit vector. The constant $-\alpha$ in the right-hand side is independent of the choice of the unit vector e, since for $|\rho| \to \infty$ the left-hand side grows without bound uniformly with respect to e.

The remainder of the reasoning can be taken over literally from the case of one space dimension.

One obtains in particular, for systems which are Petrovskiĭ-parabolic (Example 3, Section 2.1) the following theorem:

Let the initial functions $u_j(x, 0)$ of a system of order p which is Petrovskiĭ-parabolic satisfy the inequality

$$|u_j(x, 0)| \leqslant C \exp b\, |x|^{p'} \qquad \left(\frac{1}{p} + \frac{1}{p'} = 1\right).$$

Then the system admits a solution $u(x, t)$ which satisfies the inequality

$$|u(x, t)| \leqslant C_1 \exp[(b + \delta)|x|^{p'}]$$

for any $\delta > 0$ and sufficiently small $t < T(\delta)$.

This theorem applies, in particular, to the heat equation (Example 1):

$$\frac{\partial u}{\partial t} = \frac{\partial^2 u}{\partial x^2};$$

Here $p = 2$, $p' = 2$ and it follows that *for any function $u(x)$ satisfying the inequality*

$$|u(x)| < C \exp[bx^2],$$

the heat equation admits a solution $u(x, t)$ which for any $\epsilon > 0$ and sufficiently small t, admits the inequality:

$$|u(x, t)| \leqslant C_1 \exp[(b + \epsilon)x^2].$$

In fact, the results are valid also for more general systems, with coefficients depending on the space coordinates. A system

$$\frac{\partial u}{\partial t} = P\left(x, t, \frac{\partial}{\partial x}\right) u \tag{14}$$

is called parabolic in the Petrovskiĭ sense (Petrovskiĭ-parabolic), if it has the following property: under replacement of the variables x and t by constants ξ, τ the system which results is Petrovskiĭ-parabolic, as defined at the beginning of this section. Under the assumption that the coefficients are continuous and everywhere bounded, as well as their derivatives up to order p, the order of the system, S. D. Eĭdel'man has shown that the Cauchy problem belonging to the initial condition

$$u(x, 0) = u_0(x)$$

has a unique solution in the class of functions satisfying the inequality

$$|u(x, t)| \leqslant C \exp[a |x|^{p'}].$$

The solution can be written as a Poisson integral

$$u(x, t) = \int_R G(x, \xi, t) u_0(\xi)\, d\xi,$$

and the kernel $G(x, \xi, t)$ is subject to the condition

$$|G(x, \xi, t)| \leqslant C \exp[+Mt]\, t^{-n/p} \exp\left[-\frac{b\,|x - \xi|^{p'}}{1/(p-1)}\right] \quad \left(\frac{1}{p} + \frac{1}{p'} = 1\right).$$

Here M is the constant which bounds from above the absolute values of the coefficients of the equation (14) and of those derivatives which were mentioned above.

2.5. Systems of Nonpositive Genus

Let us now consider the case where the genus of the system (1) is not positive: $\mu \leqslant 0$. The following theorem holds here.

Theorem 2. *Let the initial functions* $\{u_j(x, 0)\} = u_0(x)$ *of a parabolic system with* $\mu \leqslant 0$ *belong to the class* $K_{p_1,0}$ *of functions* $f(x)$, *which for every* $\epsilon > 0$ *satisfy the inequality*

$$|f(x)| \leqslant C_\epsilon \exp[\epsilon \mid x \mid^{p_0}], \qquad p_1 = \frac{h}{h - \mu}, \tag{1}$$

then for sufficiently small $T > 0$ *and* $t \leqslant T$, *the solution of the system belongs to the same class.*

Proof. In the same manner as in the preceding section, we can restrict our attention to the case $n = 1$. The resolvent matrix $Q(s, t)$, considered as a function of the complex variable $s = \sigma + i\tau$ defined in the domain H_μ, where

$$|\tau| \leqslant K(1 + |\sigma|)^\mu, \qquad \mu \leqslant 0,$$

satisfies the inequality

$$\| Q(s, t) \| \leqslant C \exp[-a't \mid \sigma \mid^h]$$

For fixed $t > 0$ we apply the Theorem 4, Section 7, Chapter IV, Volume 2, which yields the following estimate for the matrix formed of the derivatives of order q of the matrix elements $Q_{ij}(s, t)$ for real $s = \sigma$:

$$\| Q^{(q)}(\sigma, t) \| \leqslant CB^q q^{q(1-\{\mu/h\})} \exp[-a''t \mid \sigma \mid^h]. \tag{2}$$

This means that the elements of the matrix $Q(\sigma, t)$ belong to the test function space $S^{\beta,B}$ (cf. Volume 2, Chapter IV, Section 1), consisting of infinitely differentiable functions ψ which satisfy the inequalities

$$|\sigma^k \psi^{(q)}(\sigma)| \leqslant C_k \bar{B}^q q^{q\beta} \qquad \left(k, q = 0, 1, 2,\ldots; \quad \beta = 1 - \frac{\mu}{h} \geqslant 1, \quad \bar{B} > B\right).$$

The theorem on Fourier transforms for the spaces $S^{\beta,B}$ (Volume 2, Chapter IV, Section 6) tells us that the elements of the matrix $G(x, t)$ belong to the space $S_{\beta,B}$. Consequently

$$\| G(x, t) \| \leqslant C \exp[-a \mid x \mid^{p_1}] \qquad p_1 = \frac{1}{\beta} = \frac{h}{h - \mu}.$$

Since $p_1 = h\,(h - \mu) \leqslant 1 \leqslant p_0\,(p_0 - 1)$, the class $K_{p_1,0}$ is contained in the uniqueness class of the Cauchy problem for the system under consideration. Within the uniqueness class, we can, according to Section 1, express the solution of the Cauchy problem in the convolution form

$$u(x, t) = G(x, t) * u(x, 0). \tag{3}$$

In the same manner as in Theorem 1, we must show that the convolution maps the class $K_{p_1,0}$ into itself. Consider the ordinary convolution

$$G(x, t) * u(x, 0) = \int G(\xi, t) u_0(x - \xi)\, d\xi. \tag{4}$$

On the basis of the facts that were proved for the matrix $G(x, t)$ and the assumptions about the initial functions $u_0(x)$, the integrand $G(\xi, t)\, u_0(x - \xi)$ admits for every $\delta > 0$ the majorant

$$\exp[-a \mid \xi - x \mid^{p_1} + \delta \mid \xi \mid^{p_1}].$$

Further, we make use of the following lemma.

Lemma 3. *Let $\lambda > 0$. For any $\alpha > 0$ and $\gamma > 0$ there exists a $\beta > 0$, such that for all x and ξ the inequality*

$$-\alpha \mid \xi - x \mid^{\lambda} + \beta \mid \xi \mid^{\lambda} \leqslant \gamma \mid x \mid^{\lambda} \tag{5}$$

is valid.

Proof. Set $x\,\xi = \rho$. Then (5) goes over into the equivalent inequality

$$\gamma \mid \rho \mid^{\lambda} + \alpha \mid 1 - \rho \mid^{\lambda} \geqslant \beta. \tag{6}$$

But the left-hand side is effectively bounded below for all ρ by a positive constant. Hence there exists a number β with the desired properties.

The lemma implies that for given numbers $a > 0$ and $\epsilon > 0$, there is a $\delta > 0$, such that

$$\tfrac{1}{2}a \mid \xi \mid^{p_1} + \delta \mid \xi - x \mid^{p_1} \leqslant \epsilon \mid x \mid^{p_1},$$

It follows that

$$\exp[-a \mid \xi \mid^{p_1} + \delta \mid \xi - x \mid^{p_1}] \leqslant \exp[\epsilon \mid x \mid^{p_1}] \cdot \exp[-\tfrac{1}{2}a \mid \xi \mid^{p_1}].$$

Therefore the integral

$$u(x, t) = \int G(x - \xi, t) u_0(\xi)\, d\xi \tag{7}$$

converges for any $t > 0$ and the result satisfies the inequality

$$|u(x, t)| \leqslant C_\epsilon' \exp[\epsilon |x|^{p_1}],$$

for any given $\epsilon > 0$.

Consequently the result of the convolution (7) belongs to the class $K_{p_1,0}$ for any $t > 0$.

The fact that the convolution (3) and the convolution (7) are the same (the first is an ordinary convolution, the second in the sense of generalized functions) is proved in the same manner as for Theorem 1. Thus the proof of Theorem 2 is complete.

Example 1. Consider the equation of order $p > 2$ (cf. Section 2.1)

$$\frac{\partial u}{\partial t} = \frac{\partial^2 u}{\partial x^2} + i\left(i\frac{\partial}{\partial x}\right) u. \tag{8}$$

The characteristic root $\lambda(s)$ for this equation is

$$\lambda(s) = -s^2 + is^p, \qquad \Lambda(\sigma) = \operatorname{Re}\lambda(\sigma) = -\sigma^2.$$

Thus

$$p_0 = p, \qquad h = 2.$$

The genus μ of Eq. (8) is determined on the basis of Theorem 1, Section 7, Chapter IV, Volume 2:

$$\mu = 1 - (p_0 - h) = 3 - p \leqslant 0.$$

Hence

$$p_1 = \frac{h}{h - \mu} = \frac{2}{p - 1}.$$

Making use of Theorem 2 we have: if the initial function $u_0(x)$ satisfies for each $\epsilon > 0$ the inequality

$$|u_0(x)| \leqslant C_\epsilon \exp[\epsilon |x|^{2/p-1}],$$

then the solution $u(x, t)$ of the equation (8) satisfies a similar inequality for arbitrary $t \in [0, T]$.

Example 2. Consider the system of equations with characteristic roots

$$\lambda_1(s) = is^6 - s^4, \qquad \lambda_2(s) = is^4 - s^2.$$

Here $p_0 = 6$ and $h = 2$. Then, from the Example in Sec. 2.3, the genus of the system is -1. Hence $p_1 = h/(h-\mu) = \frac{2}{3}$ and Theorem 2 yields: If the initial function $u_0(x)$ satisfies for each $\epsilon > 0$ an inequality

$$|u_0(x)| \leqslant C_\epsilon \exp[\epsilon \mid x \mid^{2/3}],$$

then the solution $u(x, t)$ satisfies for any $\epsilon > 0$ a similar inequality:

$$|u(x, t)| \leqslant C_\epsilon' \exp[\epsilon \mid x \mid^{2/3}].$$

3. Hyperbolic Systems

3.1. Definition and Examples

A system with constant coefficients

$$\frac{\partial u_j(x, t)}{\partial t} = \sum_{k=1}^{m} P_{jk}\left(i \frac{\partial}{\partial x}\right) u_k(x, t) \qquad (j = 1, 2, \ldots, m) \tag{1}$$

is said to be *hyperbolic* if the function $\Lambda(s) = \max_j \operatorname{Re} \lambda_j(s)$ has the following properties:

(a) the (power-law) degree of growth of $\Lambda(s)$ is not larger than 1:

$$\Lambda(s) \leqslant a \mid s \mid + b.$$

(b) For real $s = \sigma$, $\Lambda(s)$ is bounded:

$$\Lambda(\sigma) \leqslant C.$$

We shall prove in this section that *for hyperbolic systems and only for such systems the Cauchy problem admits a solution for any sufficiently smooth initial data, without any restrictions on their growth at infinity.*

Example 1. The first-order equation

$$\frac{\partial u}{\partial t} = a \frac{\partial u}{\partial x}$$

is characterized by the function

$$\Lambda(s) = \operatorname{Re} \lambda(s) = \operatorname{Re}(-ias)$$

which is of order 1. For real $s = \sigma$

$$\Lambda(\sigma) = \operatorname{Re}(-ia\sigma) = \sigma \cdot \operatorname{Im} a,$$

so that $\Lambda(\sigma)$ is bounded only for real a. Thus Eq. (1) is hyperbolic for real a only.

Example 2. The one-dimensional wave equation

$$\frac{\partial^2 u}{\partial t^2} = a^2 \frac{\partial^2 u}{\partial x^2}$$

(a is a real constant).

The characteristic equation is

$$\lambda^2(s) = -a^2 s^2$$

and has the roots

$$\lambda_{1,2}(s) = \pm ias.$$

Obviously, the hyperbolicity conditions (a) and (b) are verified. The same is also true for the wave equation in n-dimensional space.

Example 3. *Systems which are hyperbolic according to Petrovskiĭ* (Petrovskiĭ-hyperbolic). A system (1) is said to be hyperbolic in the Petrovskiĭ sense or Petrovskiĭ-hyperbolic if its order p is 1, the characteristic roots of the matrix $P(s)$ are purely imaginary for real $s = \sigma$, $|\sigma| = 1$, the matrix P can be diagonalized for all real σ and at the same time the absolute value of the determinant of the transformation matrix is larger than some positive constant.

For such systems I. G. Petrovskiĭ has proved the correctness of the Cauchy problem in a class of sufficiently smooth functions of arbitrary growth. On the basis of the fundamental theorem (just formulated), to be proved in this section, systems which are Petrovskiĭ-hyperbolic are also hyperbolic in the sense defined by us. However, the Petrovskiĭ-hyperbolic systems do not exhaust by far the class of hyperbolic systems: the Petrovskiĭ condition is not satisfied by systems for which the Jordan structure of the matrix $P(\sigma)$ changes with σ. For instance, the system with the matrix

$$P(s) = \begin{Vmatrix} 1 & as+1 \\ bs & 1 \end{Vmatrix} \qquad (ab < 0)$$

has the characteristic roots

$$\lambda_{1,2}(s) = 1 \pm (bs(as+1))^{1/2}$$

which are distinct for $s \neq 0$, $s \neq -1/a$; in both exceptional cases $s = 0$, $s = -1/a$ the roots coincide and the matrix $P(s)$ cannot be diagonalized, so that the system is not Petrovskiĭ-hyperbolic. On the other hand, the real parts of these roots are bounded from above for real $s = \sigma$, and for complex s they grow not faster than $C|s|$. Thus the system *is* hyperbolic in the sense of our definition.

Example 4. *Equations which are hyperbolic according to Gårding* (Gårding-hyperbolic). An equation

$$\frac{\partial^m u}{\partial t^m} = L\left(\frac{\partial}{\partial t}, \frac{\partial}{\partial x_1}, \ldots, \frac{\partial}{\partial x_n}\right) \tag{2}$$

is said to be hyperbolic in the Gårding sense—or Gårding-hyperbolic—if the polynomial in the right-hand side is of order $\leqslant m$ in all arguments, of order $< m$ in the argument $\partial/\partial t$ and if the real parts of the roots of the equation $\lambda^m = L(\lambda, i\sigma_1, \ldots, i\sigma_n)$ are bounded for all real σ. We show that the system of equations which is obtained from (2) by the substitution $u_1 = u$, $u_2 = \partial u/\partial t, \ldots, u_m = \partial^{m-1}/\partial t^{m-1}$ is hyperbolic in our sense. Indeed, as has been indicated in Section 6, Chapter II, the characteristic roots $\lambda_1(s), \ldots, \lambda_m(s)$ of the system obtained after this substitution coincide with the roots of the equation $\lambda^m = L(\lambda, is_1, \ldots, is_n)$. Since the degree of the right-hand side is $\leqslant m$, in all variables, and $< m$, in the variable λ, it follows that $|\lambda_j(s)| \leqslant C|s|$ for sufficiently large $|s|$. Further, according to the assumption, $\operatorname{Re}\lambda_j(\sigma)$ are bounded. Thus the system satisfies our hyperbolicity conditions, as asserted.

3.2. The Resolvent Matrix of a Hyperbolic System

According to the general method outlined in Section 1, we subject the system (1) in Section 3.1 to a Fourier transformation, thus obtaining the system of ordinary differential equations

$$\frac{dv(s, t)}{dt} = P(s)v(s, t). \tag{1}$$

We estimate the resolvent matrix of this system

$$Q(s, t) = e^{tP(s)}.$$

Since the real parts of the characteristic roots of the matrix $P(s)$ have (power-law) degree of growth not higher than 1, it follows that

$$\|Q(s, t)\| \leqslant C(1 + |s|)^{p(m-1)}e^{bt|s|}, \tag{2}$$

as for any system of reduced order $\leqslant 1$. Further, since $\Lambda(s)$ is bounded for $s = \sigma$, we have

$$\|Q(\sigma, t)\| \leqslant C_1(1 + |\sigma|)^{p(m-1)}. \tag{3}$$

Thus the function $Q(s, t)$ is of order 1 and type bt and for real $s = \sigma$ does not increase faster than a polynomial. We shall call the smallest (nonnegative) integer h, for which the inequality

$$\| Q(\sigma, t) \| \leqslant C(1 + |\sigma|)^h \tag{4}$$

holds, the *correctness exponent* of the hyperbolic system. We see that $h \leqslant p(m-1)$. We shall show that the number h determines the degree of smoothness required of the initial functions, in terms of which the Cauchy problem admits a correct solution.

Note. If, conversely, Eqs. (2) and (3) were true, it follows from the fundamental relation (6) in Section 6.1, Chapter II, that

$$\Lambda(s) \leqslant C_1 |s| + C_2, \qquad \Lambda(\sigma) \leqslant C,$$

i.e., the system is hyperbolic. Consequently, one might use the inequalities (2) and (3) for defining hyperbolic systems. Since all that follows is based only on these inequalities, one may take into consideration also some systems for which the coefficients are functions of t; such systems are transformed into systems of ordinary differential equations by Fourier transformation and as long as their resolvent matrices $Q(s, t)$ satisfy the inequalities (2) and (3), the method can be applied and we shall call such systems *hyperbolic*, also.

3.3. The Fundamental Theorem

Theorem 1. *If the initial functions* $u_j(x, 0)$ $(j = 1, 2, \dots, m)$ *of a hyperbolic system*

$$\frac{\partial u_j(x, t)}{\partial t} = \sum_{k=1}^{m} P_{jk}\left(i \frac{\partial}{\partial x}\right) u_k(x, t) \qquad (j = 1, \dots, m) \tag{1}$$

with correctness exponent h *admit continuous derivatives with respect to* x *up to order* $h + n + k$ *(*n *is the number of independent space variables, and* k *is a nonnegative integer), then the system admits a continuous solution* $u(x, t)$ *which is* k *times differentiable in* x.

The solution depends continuously on the initial functions $u_j(x, 0)$ *in the following sense: if the functions* $u_{j\nu}(x, 0)$ *converge for* $\nu \to \infty$, *together with their derivatives up to order* $h + k + n$, *uniformly in each ball* $x \leqslant r$ *to the function* $u_j(x, 0)$ *(and its derivatives, respectively), then the corresponding solutions* $u_{j\nu}(x, t)$ *converge to the solution* $y_j(x, t)$, *together with their derivatives in* x *up to order* k *uniformly in each ball* $|x| \leqslant r$.

Note that no restriction whatsoever has been imposed on the growth of the initial functions $u_j(x, 0)$ and their derivatives for $|x| \to \infty$.

Proof. We know that the Cauchy problem for hyperbolic systems (1) admits a unique solution within the class of generalized functions over the test function space K of infinitely differentiable functions of compact support (the space of Schwartz distributions, usually denoted by D'). In this class the solution of the Cauchy problem (1) can be written in the convolution form

$$u(x, t) = G(x, t) * u(x, 0), \tag{2}$$

where $G(x, t)$ is the inverse Fourier transform of the resolvent matrix $Q(s, t)$. It remains to be shown that (2) maps a function $u(x, 0)$ which is $h + n + k$ times differentiable into a function $u(x, t)$ admitting derivatives up to order k with respect to x. For this purpose, we consider the matrix function $Q(s, t)$. The elements of this matrix are entire analytic functions of order (at most) one and type $\leqslant bt$, which for real $s = \sigma$ increase not faster than a polynomial of degree h. We make use of the theorem in Section 4.5, Chapter III, Volume 2; according to this theorem the Fourier transform of an entire function $y(s)$ of order one and type $\leqslant \theta$, which increases for real $s = \sigma$ not faster than a polynomial of degree h, is a generalized function (i.e., a distribution, or linear functional over K), with support in the region $|x| \leqslant \theta$ and admits a representation of the form

$$F^{-1}g(s) = R\left(\frac{\partial}{\partial x}\right)f(x),$$

(F denotes the Fourier transformation); here $f(x)$ is a continuous function which vanishes outside the region $|x| \leqslant bt + \epsilon$ and R is a fixed polynomial of degree not higher than $h + n$. In the case under consideration we have for each $\theta > 0$ a T such that $bT \leqslant \theta$; then for $0 \leqslant t \leqslant T$ we have:

$$F^{-1}Q(\sigma, t) = G(x, t) = R\left(\frac{\partial}{\partial x}\right)f(x, t),$$

where $f(x, t)$ is a continuous function vanishing outside the region $|x| \leqslant \theta + \epsilon$. Hence

$$\begin{aligned} u(x, t) &= G(x, t) * u(x, 0) \\ &= R\left(\frac{\partial}{\partial x}\right)f(x, t) * u(x, 0) \\ &= f(x, t) * R\left(\frac{\partial}{\partial x}\right) u(x, 0). \end{aligned}$$

If $u(x, 0)$ admits derivatives up to order $h + k + n$, then the function $R(\partial/\partial x)\, u(x, 0)$ admits derivatives up to order k, and the integral

$$\begin{aligned} u(x, t) &= f(x, t) * R(\partial/\partial x)u(x, 0) \\ &= \int f(x - \xi, t)R(\partial/\partial \xi)u(\xi, 0)\, d\xi \end{aligned} \tag{3}$$

converges in the usual sense, since it ranges only over a bounded region $|\xi| \leqslant \theta + \epsilon$. Thus the existence of the solution has been proved under the assumed conditions. The continuity with respect to the initial conditions is also obvious from (3).

3.4. The Case $p_0 < 1$

We assert that for $p_0 < 1$ the matrix $Q(s, t)$ consists of polynomials in s.

Indeed, in this case each element $Q_{ij}(s, t)$ of the matrix $Q(s, t)$ is an entire function of order smaller than 1, which for real $s = \sigma$ increases at most as a polynomial of degree h. Then the Phragmén-Lindelöf theorem (as generalized to n variables in Section 9.2, Chapter IV) implies that $Q_{ij}(s, t)$ is a polynomial of degree not higher than h in the variables $s_1, s_2, \ldots, s_n$.

Thus if $p_0 < 1$ it follows that $p_0 = 0$. The Green's matrix $G(x, t) = F^{-1}Q(\sigma, t)$ then consists of elements of the form $P_{jk}(D, t)\,\delta(x)$, where the order of the differential operators P is at most h.

The convolution (3), Section 3.3, becomes

$$u(x, t) = P(D, t)\,\delta(x) * u_0(x) = P(D, t)u_0(x), \tag{1}$$

so that the solution of the Cauchy problem for the system under consideration is obtained by differentiating the initial functions.

As an example we can consider an arbitrary matrix $P(s)$ with constant characteristic roots. In particular we can take the matrix

$$P(s) = \left\| \begin{matrix} \sqrt{p(s)q(s)} & p(s) \\ -q(s) & -\sqrt{p(s)q(s)} \end{matrix} \right\|, \tag{2}$$

where the polynomials $p(s)$ and $q(s)$ are such that their product is the square of a polynomial. Both characteristic roots of this matrix vanish for any s.

We note a difference in the properties of solutions of the form of Eq. (3), Section 3.3 and Eq. (1) in this section. The structure of Eq. (3),

Section 3.3, shows that an excitation which is localized in a finite region propagates with a finite velocity not larger than b. On the other hand, it follows from Eq. (1) of this section, that if the initial excitation is localized in a finite region, then it will remain in that region all the time.

3.5. The Converse Theorem

In conclusion we show that hyperbolic systems are completely determined by their correctness classes.

Theorem 2. *Let it be known that a system with constant coefficients*

$$\frac{\partial u_j(x, t)}{\partial t} = \sum_{k=1}^{m} P_{jk}\left(i \frac{\partial}{\partial x}\right) u_k(x, t) \tag{1}$$

is correct in the class of all sufficiently smooth functions. This means that for arbitrary initial functions $u_0(x)$ which are h times differentiable there exists a unique solution $u(x, t)$ and for each series of initial data $u_0(x) = \sum_{\nu=1}^{\infty} u_{0\nu}(x)$, which is uniformly convergent in a bounded region, together with its derivatives up to order h, there exists a solution $u(x, t) = \sum_{\nu=1}^{\infty} u_\nu(x, t)$ with the series converging for each x and t to the corresponding solution of the system (1). *Then the system* (1) *is hyperbolic.*

Proof. We consider the function $\varphi(x) \neq 0$ which vanishes for $|x| > \frac{1}{2}$ and admits derivatives up to order h. The solution of the Cauchy problem for the system (1) with $\varphi(x)$ as initial function has the form

$$\varphi(x, t) = G(x, t) * \varphi(x),$$

where $G(x, t)$ is the Green's matrix of the system (1). Let $p_\nu > 0$ denote an arbitrary sequence of numbers and let x_ν be a sequence of points which are separated from each other by a distance not smaller than 1. By assumption of the theorem the Cauchy problem with the initial function

$$u_0(x) = \sum_{\nu=1}^{\infty} p_\nu \varphi(x + x_\nu)$$

(the series converges uniformly, together with its derivatives up to order h, in any region $|x| \leqslant M$, since only a finite number of terms does not vanish) is solved by

$$\begin{aligned} u(x, t) &= \sum p_\nu [G(x, t) * \varphi(x + x_\nu)] \\ &= \sum p_\nu (G(\xi, t), \varphi(x + \xi + x_\nu)), \end{aligned}$$

and the series converges for any x. It follows for $x = 0$ that the series

$$\sum p_\nu(G(\xi, t), \varphi(\xi + x_\nu))$$

converges for any choice of p_ν. Therefore the function

$$p(\eta)(G(x, t), \varphi(x + \eta))$$

is bounded for an arbitrary choice of the function $p(\eta)$. But this can be true only if the function

$$(G(x, t), \varphi(x + \eta))$$

vanishes for sufficiently large $|\eta|$. By passing to the Fourier transforms it follows that *the function $Q(s, t)\,\psi(s)$ is an entire function of order 1 and finite type for arbitrary functions $\psi(\sigma)$, the Fourier transform of the original function $\varphi(x)$.*

We prove that the function $Q(s, t)$ is itself of order 1.

Since $\psi(s)$ is a Fourier transform of a function of compact support $\varphi(x)$, it is a function of order 1 and finite type. This function can be so defined as to ensure that for a given value of the argument θ_j, $\theta_j \neq 0, \pi$, the quantity

$$|\psi(\sigma_1, ..., \sigma_{j-1}, r_j e^{i\theta_j}, \sigma_{j+1}, ..., \sigma_n)|$$

should go to $+\infty$ for $r \to \infty$.

One might choose, for instance,

$$\psi(s) = \sum_j \left(\frac{\sin ks_j}{ks_j}\right)^{h+2} \qquad k(h + 2) < \tfrac{1}{2}.$$

If the function $Q(s, t)$ would increase faster than a function of order 1 and finite type, there would exist a ray in the complex plane of the variable s_j with the argument θ_j, $0 < \theta_j < \pi$, for which $|Q(r_j e^{i\theta_j}, t)|$ increases faster than e^{Cr_j} for any C. But then the product $Q\psi$ would increase along this ray even faster, and thus the function $Q\psi$ could not be of first order and finite type.

Therefore we reach the conclusion that *for arbitrary t the function $Q(s, t)$ is an entire function of first order and finite type.*

One can assert further than for real $s = \sigma$ the function $Q(\sigma, t)$ does not increase faster than a polynomial in σ. Indeed, the function $(G(x, t), \varphi(x + \eta)) = \varphi(\eta, t)$ is bounded and continuous on the interval on which it does not vanish (due to the assumption of correctness). Consequently, its Fourier transform $Q(\sigma, t)\,\psi(\sigma)$ vanishes as $|\sigma| \to \infty$.

But the function $\psi(\sigma)$ can be defined so that the inequality

$$|\psi(\sigma)| \geqslant C/(1 + |\sigma|)^{h+2},$$

be satisfied, by taking, for instance,

$$\psi(\sigma) = \sum_j \left[\frac{\sin^{h+2} k\sigma_j}{\sigma_j^{h+2}} + \frac{\sin^{h+2} k(\frac{1}{2}\pi - \sigma_j)}{(\frac{1}{2}\pi - \sigma_j)^{h+2}}\right].$$

Here k should be chosen such that $k(h + 2) < \frac{1}{2}$ (in order to guarantee that the type is $<\frac{1}{2}$). Then the function $\psi(\sigma)$ will be the Fourier transform of a function $\varphi(x)$ which is novanishing only within the interval $|x| < \frac{1}{2}$ and admits derivatives up to order h. Since $\lim_{|\sigma|\to\infty} Q(\sigma, t)\, \psi(\sigma) = 0$, it follows that $\| Q(\sigma, t)\|$ increases for $|\sigma| \to \infty$ slower than $|\sigma|^{h+2}$, as required. We see that both conditions required for hyperbolicity are verified for the system (1). Consequently, under the stated conditions the system (1) is indeed hyperbolic, which concludes the proof of Theorem 2.

4. Systems Which Are Petrovskiĭ-Correct

4.1. Definition and Examples

A system with constant coefficients

$$\frac{\partial u_j(x, t)}{\partial t} = \sum_{k=1}^{m} P_{jk}\left(i \frac{\partial}{\partial x}\right) u_k(x, t) \tag{1}$$

is said to be *correct according to Petrovskiĭ* (briefly: *Petrovskiĭ-correct*) if the function $\Lambda(s) = \operatorname{Re} \lambda_j(s)$ is bounded from above for real values of $s = \sigma$:

$$\Lambda(\sigma) \leqslant C.$$

Example 1. Any parabolic system (Section 2).

Example 2. Any hyperbolic system (Section 3)

Example 3. The equation describing sound propagation in a viscous gas:

$$\frac{\partial^2 u}{\partial t^2} = 2 \frac{\partial^3 u}{\partial t\, \partial x^2} + \frac{\partial^2 u}{\partial x^2}. \tag{2}$$

Here the characteristic equation has the form

$$\lambda^2 = 2\lambda s^2 - s^2;$$

and its roots

$$\lambda_{1,2}(s) = -s^2 \pm (s^4 - s^2)^{1/2}$$

are bounded from above for real $s = \sigma$. Consequently Eq. (2) is Petrovskiĭ-correct. This equation belongs neither to the class of parabolic equations (since its roots become positive for sufficiently large $s = \sigma$), nor to the class of hyperbolic equations (since the inequality $\operatorname{Re} \lambda(s) \leqslant C|s|$ is not satisfied).

Example 4. The Schrödinger equation

$$\frac{\partial u}{\partial t} = i \frac{\partial^2 u}{\partial x^2}. \tag{3}$$

Here $\lambda(s) = -is^2$ and $\Lambda(s) = 2\sigma\tau$. Since $\Lambda(s)$ vanishes identically along the real axis, the Schrödinger equation is Petrovskiĭ-correct. This equation too, is neither parabolic nor hyperbolic.

4.2. The Resolvent Matrix

Fourier-transforming the system (1) in the previous section we obtain the system

$$\frac{dv(s, t)}{dt} = P(s)v(s, t), \tag{1}$$

with the resolvent matrix

$$Q(s, t) = e^{tP(s)}.$$

Let p_0 denote the reduced order of the system. Then we have the inequality

$$\| Q(s, t)\| \leqslant C(1 + |s|)^{p(m-1)} \exp[bt\,|s|^{p_0}]. \tag{2}$$

If $p_0 \leqslant 1$, the system (1) in the previous section is hyperbolic. We assume in this section that $p_0 > 1$, so that the system is *not* hyperbolic.

The Petrovskiĭ correctness condition implies, together with (2)

$$\| Q(\sigma, t)\| \leqslant C_1(1 + |\sigma|)^{p(m-1)}.$$

We shall again define the *correctness exponent h* of a system (1) as the smallest number h for which

$$\|Q(\sigma, t)\| \leqslant C_1(1 + |\sigma|)^h. \tag{3}$$

is verified. We shall assume the existence of h—otherwise one could have used a fixed number h arbitrarily close to the lower bound.

We see that $h \leqslant p(m - 1)$. In the same manner as for hyperbolic systems, h determines the order of smoothness (i.e., differentiability) of the initial data in order to ensure the correctness of the Cauchy problem.

Note. As in the preceding cases, the inequality (3) can serve as a definition of a system which is Petrovskiĭ-correct. Therefore one could also consider systems with variable coefficients and other systems, which after Fourier transformation take the form

$$\frac{dv(s, t)}{dt} = P(s, t)v(s, t)$$

with a resolvent matrix $Q(s, t)$ satisfying the inequalities (4)–(5). From now on all such systems will be called *Petrovskiĭ-correct*.

4.3. The Rôle of the Petrovskiĭ Correctness Condition

It will be shown in the following that for a Petrovskiĭ-correct system the existence of the solution is guaranteed for sufficiently smooth initial functions (which do not increase faster than a given order) as well as the continuity of this solution with respect to variations of the initial data.

In this section we show that the Petrovskiĭ correctness condition is necessary for the validity of similar propositions.

Theorem 1. *Let the system of equations*

$$\frac{\partial u_j(x, t)}{\partial t} = \sum P_{jk}\left(i \frac{\partial}{\partial x}\right) u_k(x, t) \tag{1}$$

admit for $-\infty < x < \infty$ *an integrable solution* $u(x, t)$ *for any initial function* $u_0(x)$ *with derivatives up to order* h. *In addition, let this solution depend continuously on* $u_0(x)$, *i.e., let*

$$\lim_{\nu \to \infty} D^q u_{0_\nu}(x) = D^q u_0(x) \qquad (|q| \leqslant h)$$

imply

$$\lim_{\nu\to\infty} u_\nu(x, t) = u(x, t)$$

for any x and t. Then the system (1) *is Petrovskiĭ-correct.*

Proof. For a function $u_0(x)$ of compact support, the solution of the Cauchy problem has the form

$$u(x, t) = G(x, t) * u_0(x), \tag{2}$$

where $G(x, t)$ is the Green's matrix of the system (1).

We replace the function $u_0(x)$ by $\epsilon u_0(x)$ with $\epsilon > 0$. Then, for $\epsilon \to 0$ the solution $\epsilon u(x, t)$ goes to zero for arbitrary x, t. This implies that the function $u(x, t)$ is finite for all x, t. Replacing $u_0(x)$ by $u_0(x + \epsilon)$ and letting ϵ go to zero, it follows that $u(x, t)$ is continuous in x for any x and t.

Applying a Fourier transformation to both sides of Eq. (2) we obtain

$$v(\sigma, t) = Q(\sigma, t)v_0(\sigma),$$

where $v(\sigma, t)$ is a bounded function (the Fourier transform of an integrable function). Taking for $v_0(\sigma)$ a function which is bounded from below by $C/(1 + |\sigma|)^h$ (cf. the end of Section 3) it follows that $Q(\sigma, t)$ increases not faster than $|\sigma|^h$. Thus the Petrovskiĭ correctness condition is satisfied, as required.

Note. Instead of the integrability of $u(x, t)$ one could have required the integrability of the ratio $u(x, t)/(1 + |x|)^k$ for some value of k. It is not known whether the theorem remains valid under the assumption that $u(x, t)$ belongs to the uniqueness class of the system (1).

4.4. The Genus of a Petrovskiĭ-Correct System

We apply the Theorem 1 of Section 7, Chapter IV, Volume 2, to the resolvent matrix function $Q(s, t)$ of the system (4), Section 4.1. This theorem implies the existence of a region H_μ defined by the inequality

$$|\tau| \leqslant K(1 + |\sigma|)^\mu, \qquad \mu \geqslant 1 - p_0, \tag{1}$$

and in which

$$\|Q(s, t)\| \leqslant C(1 + |\sigma|)^h \tag{2}$$

holds. The least upper bound of the numbers μ will be called the *genus* of the system. We shall see that the genus of the system, together with the correctness exponent h, determines the correctness class for the Cauchy problem for the given system.

In the same manner as for parabolic systems, we assume that the least upper bound is attained in the class of all admissible numbers μ, so that μ in Eq. (1) can actually be considered the genus of the system. If in reality this is not so, one can, as before, select for μ an arbitrary number which is smaller than the genus of the system and modify the final formulations accordingly.

The genus of a system with constant coefficients can be determined from its characteristic roots. On the basis of the fundamental inequality (6) in Section 6.1 of Chapter II, which relates the growth of the function $Q(s, t)$ to the growth of the function $\Lambda(s) = \max_j \operatorname{Re} \lambda_j(s)$, the genus of a system can be defined as the largest exponent such that the function $\Lambda(s)$ remains bounded in the region

$$|\tau| \leqslant K(1 + |\sigma|)^{\mu}.$$

Example 1. The sound equation (2) of Section 4.1 has the characteristic roots

$$\lambda_{1,2}(s) = -s^2 \pm (s^4 - s^2)^{1/2};$$

it is easy to verify that their real parts are bounded in any angle $|\tau| \leqslant k|\sigma|$, $0 < k < 1$. Thus the genus of this equation is 1.

Example 2. For the Schrödinger equation (3) in Section 4.1, the characteristic root is

$$\lambda(s) = -is^2, \qquad \Lambda(s) = \operatorname{Re} \lambda(s) = 2\sigma\tau,$$

and $\Lambda(s)$ is bounded in the region

$$|\tau| \leqslant K(1 + |\sigma|)^{-1};$$

thus for the Schrödinger equation the genus is -1.

It has been shown by V. M. Borok that for Petrovskiĭ-correct systems with one space variable one can construct simple formulas for the computation of all characteristic quantities of the system, in the same manner as for parabolic systems (Section 2.3). Expanding the roots of the equation $\det\| P(s) - \lambda E \| = 0$ in a series in the neighborhood of the point at infinity

$$\lambda(s) = \alpha_0 s^{k_0} + \alpha_1 s^{k_1} + \cdots + \alpha_p s^{k_p} + \cdots$$

one can distinguish three types of roots, according to their expansion:

(1) $k_0 \leqslant 0,$

(2) $k_0 > k_1 > \cdots > k_m > 0, \qquad k_{m+1} \leqslant 0,$

$\operatorname{Re} \alpha_0 = \cdots = \operatorname{Re} \alpha_m = 0,$

(3) $k_0 > k_1 > \cdots > k_m > 0, k_{m+1} \leqslant 0,$

$\operatorname{Re} \alpha_0 = \cdots = \operatorname{Re} \alpha_{p-1} = 0, \qquad \operatorname{Re} \alpha_p \neq 0, \qquad p \leqslant m.$

It turns out that for roots of type (2) the exponents $k_0, \ldots, k_m$ are integers and for roots of type (3) the exponents $k_0, \ldots, k_p$ are integers. The following formula holds

$$p_0 = \max(k_0, 0) \qquad \text{(with respect to all roots).}$$

If $p_0 = 0$, the system reduces to a system of ordinary equations (without derivatives with respect to x). Let $p_0 > 0$; then

$$\mu = \min\{-k_0 + 1, \quad k_p - k_0 + 1, \quad 1\} \qquad \text{(with respect to all roots).}$$

4.5. The Fundamental Theorem for Systems of Positive Genus

We now formulate the fundamental theorem on correctness classes for systems with positive genus:

Theorem 2. *If the initial functions $u_j(x, 0)$ of the Petrovskiĭ-correct system*

$$\frac{\partial u_j(x, t)}{\partial t} = \sum_{k=1}^{m} P_{jk}\left(i \frac{\partial}{\partial x}\right) u_k(x, t) \qquad (j = 1, \ldots, m), \tag{1}$$

of positive genus and having the correctness exponent h, satisfy the inequalities

$$| u_j^{(q)}(x, 0)| \leqslant C_1 \exp[b \mid x \mid^{p_1}], \qquad p_1 = \frac{p_0}{p_0 - \mu}, \qquad q \leqslant h + n + 1,$$

then the system admits a solution $u(x, t)$ which for sufficiently small t satisfies the inequality

$$| u_j(x, t)| \leqslant C_2 \exp[b' \mid x \mid^{p_1}]$$

for arbitrary $b' > b$.

Example. We again consider the equation describing propagation of sound in a viscous gas:

$$\frac{\partial^2 u}{\partial t^2} = 2\frac{\partial^3 u}{\partial t\,\partial x^2} + \frac{\partial^2 u}{\partial x^2} \tag{2}$$

It was shown that it is Petrovskiĭ-correct, of genus $\mu = 1$. Its characteristic roots

$$\lambda_{1,2}(s) = -s^2 \pm (s^4 - s^2)^{1/2}$$

are bounded on the real axis, thus the correctness exponent is $h = 0$. The reduced order p_0 of Eq. (2) is determined by the growth of the roots in the x-plane and is obviously equal to 2. Theorem 2 *asserts that for initial functions* $u(x, 0)$, $\partial u(x, 0)/\partial t$ *satisfying together with their derivatives up to second order inequalities of the type*

$$|u(x)| \leqslant C \exp[b\,|x|^2],$$

there exists a solution which for sufficiently small t *satisfies the inequalities*

$$|u(x, t)| \leqslant C_2' \exp[b'x^2], \; |\partial u(x, t)/\partial t| \leqslant C_2'' \exp[b'x^2]$$

with an arbitrary constant $b' > b$.

The proof of Theorem 2 necessitates several preliminary constructions. For simplicity we first carry it through for the case of one space dimension ($n = 1$).

(1). *If the function* $Q(\sigma)$ *increases for* $|\sigma| \to \infty$ *not faster than* $(1 + |\sigma|)^h$, *this function can be represented as a Fourier transform of a square-integrable function* $f(x)$ *which is acted on by a (fixed) differential operator* $P(D)$ *of order* $h + 1$ (e.g., $(1 + (d/dx))^{h+1}$).

Indeed, if we write

$$Q(\sigma) = (1 - i\sigma)^{h+1} R(\sigma),$$

then $R(\sigma)$ will be square-integrable and according to the Plancherel Theorem[6] it will have a square-integrable (inverse) Fourier transform $f(x)$:

$$\widetilde{f(x)} = R(\sigma).$$

[6] Cf. e.g., E. C. Titchmarsh, "Introduction to the Theory of Fourier Integrals," Section III, Oxford Univ. Press, 1937.

Multiplying both sides with $(1 - i\sigma)^{h+1}$, we obtain

$$Q(\sigma) = (1 - i\sigma)^{h+1}\widetilde{f(x)} = F\left[\left(1 + \frac{d}{dx}\right)^{h+1} f(x)\right],$$

as required.

(2). *Let $u(x)$ be a function of compact support with continuous derivatives up to order $h + 1$. Let $G(x)$ be a generalized function* (over a test function space to be selected later) *which admits the representation*

$$G(x) = P\left(\frac{d}{dx}\right)f(x),$$

where $P(d/dx)$ is a differential operator of order $h + 1$, and $f(x)$ is a square integrable function. We assert that the convolution $G(x) * u(x)$ can be represented in the form

$$G * u = \int f(\xi)P(d/dx)\, u(x - \xi)\, d\xi. \tag{3}$$

Indeed, for any test function $\varphi(x)$ of compact support we have, by definition of the convolution

$$\begin{aligned}
(G * u, \varphi) &= (u, G * \varphi) = (u, P(d/dx)f * \varphi) \\
&= (u, f * P(d/dx)\varphi) = \left(u, \int f(\xi)P(d/dx)\, \varphi(x + \xi)\, d\xi\right) \\
&= \int u(x) \left\{\int f(\xi)P(d/d\xi)\, \varphi(x + \xi)\, d\xi\right\} dx.
\end{aligned}$$

Both integrals are in fact over finite regions. Changing the order of integration and setting $x + \xi = \eta$, we find:

$$\begin{aligned}
(G * u, \varphi) &= \int f(\xi) \left\{\int u(\eta - \xi)P(d/d\eta)\, \varphi(\eta)\, d\eta\right\} d\xi \\
&= \int f(\xi) \left\{\int P(-d/d\eta)\, u(\eta - \xi)\, \varphi(\eta)\, d\eta\right\} d\xi \\
&= \int \left\{\int f(\xi)P(d/d\xi)\, u(\eta - \xi)\, d\xi\right\} \varphi(\eta)\, d\eta \\
&= \left(\int f(\xi)P(d/d\xi)\, u(x - \xi)\, d\xi, \varphi(x)\right).
\end{aligned}$$

This is at the same time the result of the action of the functional in the right-hand side of (3) on the test function $\varphi(x)$. Thus our assertion is proved.

(3). We denote by $L = L(h_0, l)$ the totality of functions $u(x)$ which vanish for $|x| \geqslant 1$ and which have continuous derivatives up to order h_0, satisfying the inequalities

$$|u^{(q)}(x)| \leqslant l \qquad \text{for} \qquad q \leqslant h_0 .$$

Let $Q(s)$ be an entire analytic function of $s = \sigma + i\tau$ which satisfies the inequality

$$|Q(\sigma + i\tau)| \leqslant C(1 + |\sigma|)^h e^{\Omega(b\tau)}, \tag{4}$$

where $\Omega(\tau)$ is the (downward) convex function, defining the space W^Ω (Chapter I, Section 1) and let $G(x)$ be a generalized function for which the Fourier transform coincides with $Q(s)$.

For $h_0 \geqslant h + 2$ the convolution

$$\hat{u}(x) = G(x) * u(x)$$

satisfies the inequality

$$|\hat{u}(x)| \leqslant C' l e^{-M(ax)}, \tag{5}$$

where $M(x)$ is the Young-dual of the function $\Omega(\tau)$ (Chapter I, Section 3) *and a is an arbitrary number, smaller than $1/b$.*

Proof. The function $G * u$ is the inverse Fourier transform of the product Qv, where $v(s) = \widetilde{u(x)}$. If $u(x)$ belongs to L, $v(s)$ is an entire analytic function, satisfying the inequalities

$$|s^q v(s)| \leqslant 2le^{|\tau|} \qquad \text{for} \qquad q \leqslant h_0 .$$

The product Qv satisfies the inequality

$$\begin{aligned} |Q(s)v(s)| &\leqslant 2Cl(1 + |\sigma|)^h e^{\Omega(b\tau)} e^{|\tau|} \min\{1, 1/|s|^{h_0}\} \\ &\leqslant C_1 l(1 + |\sigma|)^{h-h_0} e^{\Omega(b\tau)} e^{|\tau|} \leqslant C_1' l(1 + \sigma^2)^{-1} e^{\Omega(b'\tau)} \end{aligned}$$

for any $b' > b$. Calculating the Fourier transform of Qv by integrating along the lines $\operatorname{Im} s = \tau$, we obtain:

$$\begin{aligned} |G(x) * u(x)| &= (1/2\pi) \left| \int Q(s)v(s)e^{-isx}\, d\sigma \right| \\ &\leqslant C_2' l e^{\Omega(b'\tau)} e^{\tau x} \int \frac{d\sigma}{1 + \sigma^2} \leqslant C_3 l e^{\Omega(b'\tau) + \tau x}. \end{aligned}$$

For fixed x, we select τ to be of opposite sign to x and such that the Young inequality for the function $\Omega(b'x)$ and the Young dual $M(x/b')$ should become an equality

$$-x\tau = \Omega(b'\tau) + M(x/b').$$

We obtain thus

$$|\hat{u}(x)| = |G(x) * \varphi(x)| \leqslant C_3 l \exp[-M(x/b')]$$

Since b' is arbitrary and larger than b, $1/b'$ is an arbitrary number, smaller than $1/b$, as required.

(4). We apply the result of (3) to obtain an estimate of the solution of the Cauchy problem for the Petrovskiĭ-correct system

$$\frac{\partial u(x, t)}{\partial t} = P\left(i \frac{\partial}{\partial x}\right) u(x, t) \tag{1}$$

with an initial function $u(x)$ of compact support. The solution will be represented as a convolution

$$u(x, t) = G(x, t) * u(x),$$

with $G(x, t)$ the inverse Fourier transform of $Q(s, t)$, the resolvent matrix-function of the system

$$\frac{dv(s, t)}{dt} = P(s)v(s, t). \tag{6}$$

The assumption of Petrovskiĭ-correctness for the system (1) implies that $Q(s, t)$ is an entire analytic function of order p_0, which for real $s = \sigma$ increases not faster than $|\sigma|^h$ with some constant h. Since the genus μ of the system is positive, we can make use of Theorem 2′ of Section 7.1, Chapter IV, Volume 2, which yields the estimate

$$\|Q(s, t)\| \leqslant C(1 + |\sigma|)^h \exp[b't\,|\tau|^{p_0/\mu}] \leqslant C(1 + |\sigma|)^h \exp[\theta\,|\tau|^{p_0/\mu}], \tag{7}$$

with $\theta = bT'$, $t \leqslant T$. This estimate coincides with (4), above, if one selects $\Omega(\tau) = \tau^{p_0/\mu}/(p_0/\mu)$ and determines b from the equation

$$\Omega(b\tau) = \frac{b^{p_0/\mu}\tau^{p_0/\mu}}{p_0/\mu} = \theta\tau^{p_0/\mu}$$

We obtain

$$b = \left(\frac{p_0}{\mu}\,\theta\right)^{\mu/p_0} = cT^{\mu/p_0}.$$

The Young dual function $M(x)$ is here x^{p_1}/p_1 with $(1/p_1)+(1/(p_0/\mu)) = 1$. Hence $p_1 = p_0/(p_0 - \mu)$. Therefore the final inequality (5) of (3) takes the form

$$|\hat{u}(x)| \leqslant Cl \exp\left(-\frac{a^{p_1}|x|^{p_1}}{p_1}\right),$$

where a is an arbitrary number smaller than $(1/b) = c_1 T^{-\mu/p_0}$. Thus, *for any* $\delta > 0$ *the following inequality holds*

$$|\hat{u}(x)| \leqslant C_\delta l \exp[-c_2 T^{-p_2}|x|^{p_1}(1-\delta)] \qquad \left(p_2 = \mu \frac{p_1}{p_0}\right). \tag{8}$$

This inequality is true in particular for the solution of the Cauchy problem for the system (1), *if the initial function vanishes for* $|x| \geqslant 1$ *and has continuous derivatives up to order* $h_0 \geqslant h + 2$, *which are bounded by the number* l.

(5). Let now the initial vector-function $u(x)$ satisfy the conditions of the theorem:

$$|u_j^{(q)}(x)| \leqslant C \exp[b_0 |x|^{p_0}], \qquad p_1 = \frac{p_0}{p_0 - \mu}, \qquad q \leqslant h_0 .$$

We show that *there exists a solution* $u(x, t)$ *which goes over into* $u(x)$ *for* $t = 0$ *and satisfies the inequality*

$$|u_j(x, t)| \leqslant C' \exp[b_1 |x|^{p_1}]$$

for given $b_1 > b_0$ *and sufficiently small* $t \leqslant T$.

The idea of the proof is as follows. We represent the initial function $u(x)$ as a series of functions of compact support

$$u(x) = \sum_{-\infty}^{\infty} u_\nu(x - \nu),$$

where $u_\nu(x)$ vanishes for $|x| \geqslant 1$. For each of the functions $u_\nu(x)$ one can represent the corresponding solution $u_\nu(x, t)$ as the convolution

$$u_\nu(x, t) = G(x, t) * u_\nu(x - \nu).$$

Making use of the result of (3) above, we show that the series

$$u(x, t) = \sum_{-\infty}^{\infty} u_\nu(x, t)$$

converges and represents the solution of the problem.

We now describe this construction in detail.

Let $e(x)$ denote a function which vanishes for $|x| \geqslant \frac{3}{4}$, equals one for $|x| \leqslant \frac{1}{4}$, has everywhere values between zero and one, admits continuous derivatives up to order h_0 and is such that (partition of unity)

$$\sum_{-\infty}^{\infty} e(x-\nu) \equiv 1. \tag{9}$$

Multiplying (9) by $u(x)$ we find

$$u(x) = \sum_{-\infty}^{\infty} u(x)\, e(x-\nu) = \sum_{-\infty}^{\infty} u_\nu(x-\nu),$$

where $u_\nu(x) = u(x+\nu)\, e(x)$ is a function with continuous derivatives up to order h_0 and vanishing outside the interval $|x| \leqslant \frac{3}{4}$.

Let $K_\nu(u)$ denote the largest of the absolute values of the function $u(x)$ and of its derivatives up to order h_0 in the interval $-\nu - \frac{3}{4} \leqslant x \leqslant -\nu + \frac{3}{4}$. It is obvious that for arbitrary $\epsilon > 0$ the inequality

$$K_\nu(u) \leqslant C \exp[b_0(|\nu| + \tfrac{3}{4})^{p_1}] \leqslant C_\epsilon \exp[(b_0+\epsilon)|\nu|^{p_1}].$$

is true.

Let K be the largest of the absolute values taken on by the function $e(x)$ and its derivatives up to order h_0. Use of the Leibniz formula shows that the absolute values $|u_\nu^{(b)}(x)|$ for $q \leqslant h_0$ are bounded by

$$CKK_\nu(u) \leqslant C_\epsilon' \exp[(b_0+\epsilon)|\nu|^{p_1}].$$

Due to the result of paragraph (4), the solution of the Cauchy problem with the initial function $u_\nu(x)$ is majorized by

$$|u_\nu(x,t)| = |G(x,t) * u_\nu(x)| \leqslant C_{\epsilon\delta} \exp[(b_0+\epsilon)|\nu|^{p_1}] \exp[-c_2(1-\delta)T^{-p_2}|x|^{p_1}].$$

We construct the function

$$u(x,t) = \sum_{-\infty}^{\infty} u_\nu(x-\nu, t). \tag{10}$$

It follows from the preceding majorization that

$$|u_\nu(x-\nu, t)| \leqslant C_{\epsilon\delta} \exp[(b_0+\epsilon)|\nu|^{p_1} - c_2(1-\delta)\, T^{-p_2}|x-\nu|^{p_1}], \tag{11}$$

consequently, *for sufficiently small T, such that $c_2(1-\delta)\, T^{-p_2} > b_0 + \epsilon$, the series* (10) *converges uniformly and absolutely on each finite interval of the x-axis. Thus $u(x, t)$ is a continuous function.*

It remains only to be shown that this function is the required solution of the Cauchy problem.

(6) We show that, *for any sufficiently small* T, *the partial sums of* (10):

$$S_N(x, t) = \sum_{-N}^{N} u_\nu(x - \nu, t)$$

admit a majorant which does not depend on N:

$$| S_N(x, t)| \leqslant C_{\epsilon\delta} \exp[(b_0 + 3\epsilon) | x |^{p_1}]. \tag{12}$$

We carry out the following transformation of the exponent of the right-hand side of (11):

$$\begin{aligned}(b_0 + \epsilon) | \nu |^{p_1} &- c_2(1 - \delta) T^{-p_2} | x - \nu |^{p_1} \\ &= (b_0 + 3\epsilon)| x |^{p_1} + [(b_0 + \epsilon) | \nu |^{p_1} \\ &\quad - (b_0 + 3\epsilon) | x |^{p_1} - c_2(1 - \delta) T^{-p_2} | x - \nu |^{p_1}]\end{aligned}$$

In order to prove the inequality (12) it is sufficient to show that small enough T the following inequality holds:

$$(b_0 + \epsilon) | \nu|^{p_1} - (b_0 + 3\epsilon) | x |^{p_1} - c_2(1 - \delta) T^{-p_2} | x - \nu |^{p_2} \leqslant -\epsilon | \nu |^{p_1}, \tag{13}$$

The inequality (13) is equivalent to

$$(b_0 + 2\epsilon) | \nu |^{p_1} \leqslant (b_0 + 3\epsilon) | x |^{p_1} + c_2(1 - \delta) T^{-p_2} | x - \nu |^{p_1}$$

or, after the substitution $\xi = \nu/x$, to the inequality

$$(b_0 + 2\epsilon) | \xi |^{p_1} \leqslant b_0 + 3\epsilon + c_2(1 - \delta) T^{-p_1} | 1 - \xi |^{p_1}. \tag{14}$$

Since the coefficient $c_2(1 - \delta) T^{-p_2}$ is unbounded for $T \to 0$, it is clear that the inequality (14) is indeed true for sufficiently small T. At the same time (13) is verified, and hence also (12). Thus, *the functions* $S_N(x, t)$, *and also* $u(x, t)$, *the limit of the sequence* $S_N(x, t)$, *are majorized by the function* $\exp[(b_0 + 3\epsilon)| x |^{p_1}]$.

(7) We consider the space W_{p_1,b_1} of infinitely differentiable functions $\varphi(x)$ satisfying the inequalities (for the time being, b_1 is arbitrary)

$$| \varphi^{(q)}(x)| \leqslant C_{\delta q} \exp\left[-\frac{1}{p_1} [(b_1 - \delta) | x |]^{p_1}\right] \qquad (\delta > 0 \text{ arbitrary}).$$

On the basis of the estimate (7) and of Theorem 1 in Section 2.4, Chapter I, the matrix $Q(s, t)$ is a bounded multiplier in the Fourier dual

space $W^{p_1', 1/b_1}$ (Section 3, Chapter I) and consequently $G(x, t)$ is a convolutor in the space W_{p_1, b_1} (i.e., a generalized function for which the convolution with an element of that space is again an element of the space). We select the number $(1/p_1)\, b_1^{p_1}$ to be larger than b, and ϵ so small that $b + 3\epsilon < (1/p_1)\, b_1^{p_1}$ Then the functions $S_N(x, t)$ and $u(x, t)$ which occur in (6) define continuous linear functionals on the space W_{p_1, b_1}.
Since the convolution is a continuous operation

$$G * u = G * \sum_{-\infty}^{\infty} u_\nu(x - \nu) = \sum_{-\infty}^{\infty} G * u_\nu(x - \nu) = \sum_{-\infty}^{\infty} u_\nu(x + \nu, t) = u(x, t);$$

(the equality is in the sense of the dual space W'_{p_1, b_1}). Thus, *the solution of the Cauchy problem for the system* (1) *with an initial function* $u(x)$ *is the same as the function* $u(x, t)$.

Note. It follows from (3) and (10) that the final formula for the solution of the Cauchy problem under consideration, has the form

$$u(x, t) = \sum_{-\infty}^{\infty} \int_{-\infty}^{\infty} f(\xi, t) P(\partial/\partial x) u_\nu(x - \nu - \xi)\, d\xi$$

$$= \sum_{-\infty}^{\infty} \int_{-\infty}^{\infty} f(\xi, t) P(\partial/\partial x) u(x - \xi) e(x - \xi - \nu)\, d\xi.$$

If it would be legitimate to interchange the order of the signs $\sum_{-\infty}^{\infty}$ and $\int_{-\infty}^{\infty}$, we would obtain the simpler formula

$$u(x, t) = \int_{-\infty}^{\infty} f(\xi, t) P(\partial/\partial x) u(x - \xi)\, d\xi.$$

But since we do not know anything about the exponential decrease of the function $f(\xi, t)$ for $|\xi| \to \infty$, the legitimacy of such an interchange is unclear.[7]

(8) We indicate the changes one has to make in the proof in order to go over to the case of n independent variables. In this case the matrix $Q(s, t)$ satisfies the inequality

$$\| Q(s, t)\| = \| Q(s_1, \ldots, s_n, t)\|$$
$$\leqslant C(1 + |\sigma|)^h \exp[\Omega(b\tau_1) + \cdots + \Omega(b\tau_n)] \qquad (\Omega(\tau) = \tau^{p_0}).$$

[7] This problem has been solved since the appearance of the Russian edition; cf. G. I. Eskin, Obobshchenie teoremy Paley-Wiener'a-Schwartz'a (A Generalization of the Paley-Wiener-Schwartz Theorem), *Usp. Mat. Nauk* **16**, Nr. 1, 185–188 (1961).

The proposition (1) remains true if one replaces the operator $(1 + d/dx)^{h+1}$ by

$$\left(1 + \frac{\partial}{\partial x_1}\right)^{h+1} \left(1 + \frac{\partial}{\partial x_2}\right) \cdots \left(1 + \frac{\partial}{\partial x_n}\right)$$

which is an operator of order $n + h$. Correspondingly, in (2) one has to require that the function $u(x) = u(x_1, \ldots, x_n)$ be continuously differentiable up to order $h + n$.

In (3) the function $u(x) = u(x_1, \ldots, x_n)$ should be required to have continuous derivatives up to order $h_0 \geqslant h + n + 1$, and bounded by the number l.

In (4), instead of making use of the Theorem 2′ from Section 7, Chapter IV, Volume 2, one should use its n-dimensional analog in Section 9 of the same Chapter.

In (5) the index ν should be replaced by the n-tuple $(\nu_1, \ldots, \nu_n)$, $-\infty < \nu_j < \infty$. The function $e(x) = e(x_1, \ldots, x_n)$ is constructed as a product $e(x_1) \cdots e(x_n)$ of one-dimensional functions $e(x_j)$ as defined in (5). Such a function $e(x)$ vanishes outside the n-dimensional ball of radius $\frac{3}{4}n^{1/2}$ and satisfies the condition (partition of unity)

$$\sum_{\nu} e(x - \nu) = \sum_{\nu_1, \ldots, \nu_n} e_1(x_1 - \nu_1) \cdots e_n(x_n - \nu_n) \equiv \prod_{j=1}^{n} \sum_{\nu_j=-\infty}^{\infty} e_j(x_j - \nu_j) \equiv 1.$$

The other parts of the proof remain unchanged.

4.6. The Case of a System of Nonpositive Genus

We now consider the case $\mu \leqslant 0$. In this case the solution of the Cauchy problem will be correct if the initial functions and their derivatives up to a certain order satisfy the condition of *increasing according to a power law.*

Consider a Petrovskiĭ-correct system

$$\frac{\partial u(x, t)}{\partial t} = P\left(i \frac{\partial}{\partial x}\right) u(x, t) \tag{1}$$

with the initial condition

$$u(x, 0) = u(x). \tag{2}$$

According to Theorem 4′ of Section 7, Chapter IV, Volume 2, the derivatives of the resolvent matrix $Q(s, t)$ admit, for $\mu \leqslant 0$, the estimate

$$\| D^q Q(\sigma, t) \| \leqslant C_q(1 + |\sigma|^{r_q}),$$

with $|r_q| \leqslant h - \mu |q|$ (here h is the correctness exponent).

Theorem 3. *For any* $l > 0$ *the Cauchy problem* (1) *and* (2) *is correct in the class of functions* $u(x)$ *which have a power-law increase of degree not higher than* $l - (n + 1)$ *together with their derivatives up to order* $r + n + 1$:

$$|D^q u(x)| \leqslant C(1 + |x|)^{l-(n+1)} \qquad (|q| \leqslant r_l + n + 1).$$

The *solution of the Cauchy problem* (1) *and* (2) *is also a function of power-law growth of degree* $\leqslant l$.

The *proof* will again be split into several parts. We consider first the case of one independent variable x ($n = 1$).

(1). Similar to the proof of Theorem 2, we introduce the class $L = L(h_0, l)$ of functions $u(x)$ which vanish for $|x| \geqslant 1$ and have continuous derivatives up to order h_0 which are bounded as follows:

$$|u^{(q)}(x)| \leqslant l \qquad \text{for} \qquad q \leqslant h_0.$$

Let $Q(\sigma)$ be a function admitting derivatives up to order k and satisfying the inequalities

$$|Q^{(q)}(\sigma)| \leqslant C(1 + |\sigma|)^r \qquad (q = 0, 1, \ldots, k).$$

Let further $G(x)$ denote a generalized function for which the Fourier transform coincides with $Q(\sigma)$.

For $h_0 \geqslant r + 2$ *the convolution*

$$\hat{u}(x) = G(x) * u(x)$$

satisfies the inequality

$$|\hat{u}(x)| \leqslant C'l \frac{1}{(1 + |x|)^k}.$$

For the proof, we note that the function $\hat{u}(x)$ is the (inverse) Fourier transform of the function $Q(\sigma)\, v(\sigma)$, where $v(\sigma)$ is the Fourier transform of $u(x)$. The function $v(\sigma)$ satisfies the inequality

$$|\sigma^r v^{(q)}(\sigma)| = \left| \int [x^q u(x)]^r \, dx \right| \leqslant C_1 l \qquad (q = 0, 1, \ldots, k; \quad r = 0, 1, \ldots, h_0)$$

and consequently

$$|v^{(q)}(\sigma)| \leqslant \frac{C_2 l}{(1 + |\sigma|)^{h_0}} \qquad (q = 0, 1, \ldots, k).$$

Furthermore the product $Q(\sigma)\, v(\sigma)$ satisfies the inequalities

$$|[Q(\sigma)v(\sigma)]^k| \leqslant \sum C_k^q \,|Q^{(q)}(\sigma)v^{(k-q)}(\sigma)|$$

$$\leqslant \sum C_k^q \cdot C(1+|\sigma|)^r \cdot \frac{C_2 l}{(1+|\sigma|)^{h_0}} \leqslant C_3 l(1+|\sigma|)^{r-h_0}.$$

It follows that for $h_0 \geqslant r+2$

$$|x^k[G(x) * u(x)]| \leqslant \int |[Q(\sigma)v(\sigma)^{(k)}]|\, d\sigma \leqslant C_3 l \int \frac{d\sigma}{(1+|\sigma|)^2} \leqslant C_4 l.$$

Thus, for $h_0 \geqslant r+2$

$$|\hat{u}(x)| \leqslant \frac{C' l}{(1+|x|)^k},$$

as required.

(2) The result can be applied for estimating the solution of the Cauchy problem for the Petrovskiĭ-correct system

$$\frac{\partial u(x, t)}{\partial t} = P\left(i \frac{\partial}{\partial x}\right) u(x, t) \tag{1}$$

with an initial function of compact support $u(x)$. The solution can be written, as always, as a convolution of $u(x)$ with the Green's matrix of the system (1)

$$u(x, t) = G(x, t) * u(x). \tag{2}$$

The Fourier transform of $G(x, t)$ is the resolvent matrix $Q(s, t)$ of the system

$$\frac{dv(s, t)}{dt} = P(s)v(s, t). \tag{3}$$

Since the system (1) is Petrovskiĭ-correct, the matrix function $Q(s, t)$ does not grow for real $s = \sigma$ faster than $|\sigma|^h$ with some h.

The genus of the system (1) was assumed nonpositive. The use of Theorem 4′ in Section 7.6, Chapter IV, Volume 2 leads to the estimate

$$\| Q^{(q)}(\sigma, t)\| \leqslant C_q(1+|\sigma|)^{h-\mu q} \qquad (q = 0, 1, \ldots).$$

Let r_k denote the smallest of the numbers for which

$$\| Q^{(q)}(\sigma, t)\| \leqslant C(1+|\sigma|)^{r_k} \qquad (q = 0, 1, \ldots, k).$$

We further assume that the initial function $u(x)$ vanishes for $|x| \geqslant 1$ and has continuous derivatives up to order h_0 which are bounded in absolute value by l. Then, as was shown above, *for* $h_0 \geqslant r_k + 2$ *the solution* $u(x, t)$ *satisfies the etimate*

$$|u_j(x, t)| \leqslant C'l \frac{1}{(1 + |x|)^k}.$$

(3) Assume now that the initial function satisfies the conditions of the theorem

$$|u_j^{(q)}(x)| \leqslant C(1 + |x|)^{l-2} \qquad (q = 0, 1, \ldots, r_j + 2). \tag{4}$$

We show that *there exists a solution of the Cauchy problem* (1) *and* (2) *which also has a power law behavior as* $x \to \infty$. For simplicity we assume again that we are dealing with a single equation.

Consider the function $e(x)$ defined in Section 4.5 (paragraph (5)): for $|x| \geqslant \frac{3}{4}$, $e(x)$ vanishes, for $|x| \leqslant \frac{1}{4}$, $e(x) = 1$, its values are everywhere between zero and one, admits derivatives up to order $h_0 = r_\nu + 1$ and satisfies the condition

$$\sum_{\nu=-\infty}^{\infty} e(x - \nu) \equiv 1.$$

In the same manner as in the proof of Theorem 2, we have

$$u(x) = \sum u(x)e(x - \nu) = \sum u_\nu(x - \nu),$$

with $u(x) = u(x + \nu)\, e(x)$ possessing continuous derivatives up to order h_0, which vanish identically outside the interval $|x| \leqslant \frac{3}{4}$. The Leibniz formula yields for the derivatives of $u_\nu(x)$

$$|u_\nu^{(q)}(x)| \leqslant \sum C_q^s\, |u^{(s)}(x + \nu)|\, K \leqslant C_1(1 + |\nu| + \tfrac{3}{4})^{l-2}$$
$$(q = 0, 1, \ldots, r_q + 1).$$

It was proved above that this implies for the solution $u_\nu(x, t)$ of the Cauchy problem with initial function $u_\nu(x)$ the inequality

$$|u_\nu(x, t)| \leqslant C' \frac{(|\nu| + \frac{7}{4})^{l-2}}{(1 + |x|)^l}.$$

We now construct the function

$$u(x, t) = \sum u_\nu(x - \nu, t). \tag{5}$$

Since

$$|u_\nu(x-\nu, t)| \leqslant C' \frac{(|\nu| + \frac{7}{4})^{l-2}}{(1+|x-\nu|)^l}. \tag{6}$$

the series converges absolutely and uniformly on each closed interval and represents a continuous function.

It remains to be shown that $u(x, t)$ is the solution we were looking for.

For this we first estimate the terms of the series (5) in order to find a common majorant for the partial sums. We have

$$|\nu| \leqslant |\nu - x| + |x|.$$

It follows from here that

$$1 + |\nu| \leqslant 1 + |x| + |x - \nu| \leqslant (1 + |x|)(1 + |x - \nu|)$$

and

$$\frac{1}{1 + |x - \nu|} \leqslant \frac{1 + |x|}{1 + |\nu|}. \tag{7}$$

Substituting (7) in (6) we obtain

$$\begin{aligned} |u_\nu(x-\nu, t)| &\leqslant C' \frac{(\frac{7}{4} + |\nu|)^{l-2}(1+|x|)^l}{(1+|\nu|)^l} \\ &\leqslant C'' \frac{(1+|x|)^l}{1+\nu^2}. \end{aligned} \tag{8}$$

i.e., the partial sums of (5), and therefore also their limit, admit the majorant

$$C''(1 + |x|)^l.$$

Consequently the function $u(x, t)$ increases according to a power law and defines a regular functional over the space S. The series (5) converges in the topology of this space.

Since all derivatives of the matrix $Q(s, t)$ increase at most according to a power law as $|\sigma| \to \infty$, this matrix is a well defined multiplier on the space S, so that its inverse Fourier transform $G(x, t)$ is a convolutor in S. Due to the continuity of the convolution

$$G * u = G * \sum u_\nu(x) = \sum G(x, t) * u_\nu(x) = \sum u_\nu(x, t) = u(x, t)$$

Consequently the convolution $G * u$ coincides with the function $u(x, t)$.

This concludes the proof for the case of a single space-variable.

For the case of n variables one has to make the following changes.

In (1) (and sequel) the inequality $h_0 \geqslant r + 2$ is to be replaced by $h_0 \geqslant r + n + 1$.

In (2) the Theorem 4′ of Section 7, Chapter IV, Volume 2 is to be replaced by its n-dimensional analog (Section 9, same Chapter). This yields the estimate

$$\| D^q Q(\sigma, t) \| \leqslant C_q (1 + |\sigma|)^{h - \mu |q|}.$$

In (3) one has to consider a function of the form $e(x) = e(x_1) e(x_2) \cdots e(x_n)$ as in the proof of Theorem 3. The condition (4) is to be replaced by the condition that the initial functions $u(x)$ satisfy the inequalities

$$| D^q u_j(x) | \leqslant C(1 + |x|)^{l-(n+1)}, \qquad |q| \leqslant r_l + n + 1.$$

The solution $u(x, t)$ has a power law order of increase $\leqslant l$.

This completes the proof of Theorem 3.

As an example let us consider the Schrödinger equation

$$\frac{\partial u(x, t)}{\partial t} = i \frac{\partial^2 u(x, t)}{\partial x^2}.$$

Here $Q(\sigma, t) \equiv \exp[-i\sigma^2 t]$, $r_q = q$, $n = 1$; for any $l > 0$ the Cauchy problem is correct in the class of functions $u(x)$ which have a power-law growth of degree $\leqslant l - 2$ together with their derivatives up to order $l + 2$.

4.7. The Converse Theorem

Since the theorem we have obtained yields a much smaller uniqueness class than for the preceding case—i.e., a power-law increase instead of exponential increase—the problem naturally arises as to whether this is the best result possible. It turns out that this is so: for $\mu \leqslant 0$, if the correctness class contains all sufficiently smooth functions of a given type of growth, the growth type is in general not faster than a power law.

More precisely this fact is expressed in the following theorem (we restrict ourselves to the case $n = 1$):

Theorem 4. *Let the resolvent matrix $Q(\sigma, t)$ satisfy inequalities of the form*

$$\| Q^{(q)}(\sigma, t) \| \leqslant C_q (1 + |\sigma|)^{r_q}, \tag{1}$$

where the numbers r_q cannot be made smaller, and

$$r_0 \leqslant r_1 \leqslant \cdots \leqslant r_k \leqslant \cdots. \tag{2}$$

Further, let the Cauchy problem be correct for the system under consideration within the class of all functions which have along the axis $-\infty < x < \infty$ *a power-law growth of degree not larger than* l, *together with their derivatives up to order* m. *Then* $m \geqslant r_{l-2} - 2$.

Before giving the proof, we consider as an example the Schrödinger equation

$$\frac{\partial u}{\partial t} = i \frac{\partial^2 u}{\partial x^2}.$$

Here the Fourier-transformed equation is

$$\frac{dv}{dt} = -is^2 v$$

and the resolvent function $Q(s, t) = \exp[-is^2 t]$ is bounded along the real axis. It is easy to see that

$$\| Q^{(q)}(\sigma, t) \| \leqslant C_q(1 + |\sigma|)^q, \qquad q = 0, 1, \ldots .$$

Thus in this case $r_q = q$. Theorem 4 claims that if the Cauchy problem for the Schrödinger equation is correct within the class of all functions with a power-law growth of degree $\leqslant l$ with derivatives up to order m, then $m \geqslant l - 4$. In other words, within the class of functions having a power law growth of degree $\leqslant l$ with derivatives up to order $\leqslant l - 3$, the Cauchy problem is manifestly incorrect, for any l. It is useful to compare this result with the one established at the end of the preceding section: within the class of all functions with power-law increase of order $\leqslant l$, together with their derivatives of order $l + 4$, the Cauchy problem is certainly correct.

It is not known whether the Cauchy problem for the Schrödinger equation is correct or incorrect for $l - 4 < m < l + 4$.

We now come to the proof of Theorem 4.

Proof. From the assumption of correctness of the Cauchy problem in a class of functions with power-law growth follows, in particular, the following: Let $u(x)$ be a function vanishing outside $|x| \geqslant 1$, possessing derivatives up to order m and let $\{x_\nu\}$ be a sequence such that $x_\nu \to \infty$, $|x_{\nu+1} - x_\nu| \geqslant 2$, then the solution of the Cauchy problem with the initial function

$$u(x) = \sum_{\nu=1}^{\infty} |x_\nu|^l \, u(x + x_\nu)$$

is represented by the formula

$$u(x, t) = \sum_1^\infty | x_\nu |^l u(x + x_\nu, t)$$

$$= \sum_1^\infty | x_\nu |^l (G(\xi, t), u(x - \xi + x_\nu)),$$

with the series converging for every x. In particular, for $x = 0$, the series

$$\sum_1^\infty | x_\nu |^l (G(\xi, t), u(x_\nu - \xi))$$

must converge. It follows that for any function $u(x)$ of the indicated type the expression $\eta^l(G(\xi, t), u(\eta - \xi))$ must vanish as $\eta \to \infty$; in particular, the following inequality must hold:

$$| \eta^{l-2}(G(\xi, t), u(\eta - \xi))| \leqslant \frac{C}{1 + \eta^2} .$$

Taking the Fourier transform, we obtain:

$$\frac{\partial^{l-2}}{\partial \sigma^{l-2}} (Q(\sigma, t) \cdot v(\sigma)) \leqslant C \tag{3}$$

for any function $v(\sigma)$ which is the Fourier transform of a function $u(x)$ of the type described above. The functions $v(\sigma)$ admit analytic continuations into the complex $s = \sigma + i\tau$ plane as entire functions of order 1 and type 1; on the other hand any entire function of order 1 and type 1 which remains absolutely integrable when multiplied by $1, \sigma, \ldots, \sigma^m$ is a Fourier transform of a function $u(x)$ of the indicated class.

The function

$$v_0(\sigma) = \frac{\sin^{m+2} \dfrac{i + \sigma}{m + 2}}{(i + \sigma)^{m+2}} ,$$

for instance, satisfies all these conditions. Therefore this function is the Fourier transform of some function $u_0(x)$ which vanishes for $| x | \geqslant 1$ and admits continuous derivatives up to order m. According to what was proved above we have:

$$\left| \frac{\partial^{l-2}}{\partial \sigma^{l-2}} (Q(\sigma, t) \cdot v_0(\sigma)) \right| \leqslant C. \tag{4}$$

But on the other hand,

$$\frac{\partial^{l-2}}{\partial \sigma^{l-2}}(Qv_0) = Q^{(l-2)}v_0 + C_1 Q^{(l-3)}v_0' + \cdots .$$

In the first term the first factor increase with a power r_{l-2} and the second factor decreases as a power $m + 2$.

In general one obtains a power-law behavior with exponent $r_{l-2} - (m + 2)$. The succeeding terms have a slower rate of increase. But the inequality (3) shows that the quantity which is obtained is in fact bounded. It follows then that $r_{l-2} \leqslant m + 2$. Thus, as asserted $m \geqslant r_{l-2} - 2$.

5. On the Solutions of Incorrect Systems

5.1. Introduction

A system of partial differential equations

$$\frac{\partial u_j(x, t)}{\partial t} = \sum_{k=1}^{m} P_{jk}\left(i \frac{\partial}{\partial x}\right) u_k(x, t) \tag{1}$$

is said to be *incorrect*, if the function $\Lambda(s) = \max_j \operatorname{Re} \lambda_j(s)$ increases according to a power law for real $s = \sigma$; i.e., for at least one sequence $\sigma_n \to \infty$

$$\Lambda(\sigma_n) \geqslant C \mid \sigma_n \mid^p, \qquad C > 0, \quad p > 0. \tag{2}$$

In this case the matrix function $Q(s, t)$ increases along the real axis faster than any power of $|\sigma|$. As was shown in Section 4, the Cauchy problem for the system (1) will then not be correct for initial functions with a finite order of smoothness. One might hope to obtain a solution (in the form of a function) which depends continuously on the initial data only if one requires the initial functions to be infinitely smooth (of class C^∞). Here an essential role will be played by upper bounds on the function $\Lambda(\sigma)$. Indeed, if the inequality

$$\Lambda(\sigma) < C \mid \sigma \mid^h + C_1$$

holds with $h < 1$, there exist solutions for infinitely differentiable initial functions with some restrictions on the growth of the derivatives. These restictions will be the stronger, the larger h is. If h becomes unity or larger, these restrictions lead to the requirement that the initial functions be analytic. Any further increase of h leads to even stronger restrictions on the growth of the initial functions in the complex $x + iy$ plane.

5.2. Conditionally Correct Systems

A system of equations (1), Section 5.1, is said to be *conditionally correct* if the function $\Lambda(s) = \max_j \operatorname{Re} \lambda_j(s)$ satisfies an inequality

$$\Lambda(\sigma) \leqslant C \mid \sigma \mid^h + C_1, \qquad h < 1.$$

Example. Consider the equation

$$\frac{\partial^2 u}{\partial t^2} = ia \frac{\partial u}{\partial x}.$$

The characteristic equation is

$$\lambda^2 = as;$$

with the roots

$$\lambda_{1,2} = \pm (as)^{1/2}.$$

On the real axis ($s = \sigma$) the function $\Lambda(s)$ increases as $(\mid \sigma \mid)^{1/2}$ (at least on one of the semiaxes). Therefore the equation belongs to the class of conditionally correct equations.

As in the preceding cases, one has to find an answer to the question: when does the convolution

$$G(x, t) * u_0(x)$$

lead to an ordinary function? Here, as usual, $G(x, t)$ is the inverse Fourier transform of the resolvent matrix-function $Q(s, t)$ of the Fourier-transform of the system (1), Section 5.1.

We analyze the properties of the function $G(x, t)$, first for the case of one independent variable.

Theorem 7 of Section 7, Chapter IV, Volume 2, implies that there exists a region H, determined by the inequality

$$\mid \tau \mid \leqslant K(1 + \mid \sigma \mid)^\mu,$$

in which the function $\Lambda(s)$ satisfies the inequality

$$\Lambda(s) \leqslant C_2 \mid \sigma \mid^h + C_3.$$

In H the function $Q(s, t)$ is bounded by

$$\|Q(s, t)\| \leqslant C_4 \exp[b \mid s \mid^h].$$

We are interested in the maximal possible value of μ. If $h < p_0$, μ cannot exceed 1. Otherwise the function $\Lambda(s)$ would have a power-law growth of degree $\leqslant h$ on each ray in the s-plane, with the exception of $\sigma = 0$, and the function $Q(s, t)$ would have an exponential growth of order $\leqslant h$, which is impossible for an entire function of order p_0.

For conditionally correct systems of positive genus $\mu > 0$, the following theorem is true:

Theorem 1. *For a conditionally correct system*

$$\frac{\partial u(x, t)}{\partial t} = P\left(i \frac{\partial}{\partial x}\right) u(x, t) \tag{1}$$

with given parameters $h, \mu > 0$, p_0 and $p_1 = p_0/(p_0 - \mu)$, the Cauchy problem with initial data

$$u(x, 0) = u(x) \tag{2}$$

is solvable for any function $u(x)$ which satisfies the conditions

$$\begin{gathered} | u^{(q)}(x)| \leqslant CA^q q^{q\alpha_1} \exp[a \mid x \mid^{\bar{p}}] \\ \left(1 < \alpha_1 < \frac{1}{h};\quad q = 0, 1, 2, \ldots;\quad \bar{p} < p_1\right). \end{gathered} \tag{3}$$

Proof. In the space $S_\beta^{\alpha_1}(\beta > 1 - \mu/p_0)$ one can construct a function $e(x)$ which vanishes for $|x| > \frac{3}{4}$, equals 1 for $|x| \leqslant \frac{1}{4}$, has values between 0 and 1 everywhere, and satisfies the condition

$$\sum_{-\infty}^{\infty} e(x - \nu) \equiv 1.$$

By means of this "partition of unity" one can decompose the initial function $u(x)$ into a series of functions of compact support

$$u(x) = \sum u(x)e(x - \nu) = \sum_{-\infty}^{\infty} u_\nu(x - \nu),$$

where

$$u_\nu(x) = u(x + \nu)e(x).$$

The derivative of order q of $u_\nu(x)$ has the estimate

$$|u_\nu^{(q)}(x)| \leqslant \sum C_q^j\, |u^{(j)}(x+\nu)e^{(q-j)}(x)|$$

$$\leqslant \sum_{j=0}^{q} C_q^j \cdot C \cdot A^j j^{j\alpha_1} \exp[q(1+|\nu|)^{\bar{p}}] C' A_1^{q-j} (q-j)^{(q-j)\alpha_1}$$

$$\leqslant C'' A_2{}^q q^{q\alpha_1} \exp[a(1+|\nu|)^{\bar{p}}].$$

Thus the function $u_\nu(x)$ belongs to the space $S_{0,1}^{\alpha_1}$. The solution of the Cauchy problem with initial function $u_\nu(x)$ is, as usual, representable in the form $u_\nu(x, t) = G(x, t) * u_\nu(x)$, where $G(x, t)$ is the Green's matrix of the system (1). We show that $u_\nu(x, t)$ is an ordinary function and that the series

$$u(x, t) = \sum_{-\infty}^{\infty} u_\nu(x-\nu, t),$$

converges and represents the required solution of the Cauchy problem.

For the proof we consider the expression

$$G(x, t) * \varphi(x), \qquad \varphi(x) \in S_{0,1}^{\alpha_1}.$$

which is meaningful, since $G(x, t)$ is a convolutor in the space $S_\beta^{\alpha_1} \supset S_0^{\alpha_1}$, which in turn follows from the fact that $Q(\sigma, t)$ is a multiplier in the space $S_{\alpha_1}^{\beta}$ ($\beta = 1/p_1$). The result of the convolution is a function belonging to the space $S_\beta^{\alpha_1}$, which we are going to estimate. Since the convolutor $G(x, t)$ is a bounded operator, it maps any bounded set in $S_\beta^{\alpha_1}$ into another bounded set. In particular the set of functions $\varphi(x)$ satisfying the inequalities

$$|\varphi^{(q)}(x)| \leqslant C B^q q^{q\alpha_1}$$

and vanishing outside a fixed interval, is bounded in $S_\beta^{\alpha_1}$. Consequently the functions

$$\psi(x) = G(x, t) * \varphi(x)$$

will satisfy the inequalities

$$|\psi(x)| \leqslant C_1 \exp[-b\,|x|^{p_1}]$$

with fixed C_1 and b. Here the constant C_1 can be considered proportional to C. With $\varphi(x) = u_\nu(x)$, the constant B is fixed ($= A_2$) and $C = C'' e^{a(1+|\nu|)^{\bar{p}}}$. Therefore

$$|G(x, t) * u_\nu(x)| \leqslant C_2 \exp[a(1+|\nu|)^{\bar{p}} \cdot e^{-b|x|^{p_1}}].$$

Further

$$|G(x, t) * u_\nu(x - \nu)| \leqslant C_2 \exp[a(1 + |\nu|)^{\bar{p}}] \exp[-b|x|^{p_1}]. \tag{4}$$

Obviously, the series with terms (4) converges absolutely and uniformly on any finite interval.

In the same manner as in Section 4, it is easy to check that the partial sums of this series have a common majorant of the form

$$C \exp[a_1 |x|^{p_1}].$$

Consequently, the series converges in the sense of generalized functions over the space $S_\beta^{\alpha_1}$. Since the convolution operator is continuous, it follows that

$$\begin{aligned} G(x, t) * u(x) &= G(x, t) * \sum u_\nu(x - \nu) \\ &= \sum G(x, t) * u_\nu(x - \nu) = u(x, t). \end{aligned}$$

Thus, as asserted, the solution $G(x, t) * u(x)$ is an ordinary function.

Let now $\mu \leqslant 0$. Then, according to Section 7, Chapter IV, Volume 2, the matrix $Q(\sigma, t)$ satisfies the inequalities

$$\| Q^{(q)}(\sigma, t) \| \leqslant C q^{q(1-\mu/h)} \exp[b|\sigma|^h] \qquad (q = 0, 1, 2, \ldots).$$

The remainder of the reasoning is similar to the preceding one, with the replacement of p_1 by h. As a result we obtain the theorem:

Theorem 2. *In the conditionally correct case, for* $h < 1$, $\mu \leqslant 0$, *the correctness class for the Cauchy problem* (1) *and* (2) *is the class of functions* $u(x)$ *which satisfy inequalities of the form*

$$|u^{(q)}(x)| \leqslant C A^q q^{\alpha_1 q} \exp[b|x|^{h/(h-\mu)}] \qquad (q = 0, 1, 2, \ldots).$$

The Schrödinger equation can be considered to belong to the conditionally correct case, taking $p_0 = 2$, $-1 < \mu < 0$, $h = 1 + \mu$; the correctness class then consists of the infinitely differentiable functions which staisfy the inequalities

$$|u^{(q)}(x)| \leqslant C A^q q^{q/(1+\mu)} \exp[b|x|^{1+\mu}].$$

One obtains a system of correctness classes, which are not contained in each other.

5.3. Correctness in the Domain of Analytic Functions

In this section, we shall clarify the problem of correctness classes of the Cauchy problem for the system

$$\frac{\partial u_j(x, t)}{\partial t} = \sum_{k=1}^{m} P_{jk}\left(i\frac{\partial}{\partial x}\right) u_k(x, t) \qquad (j = 1, \ldots, m) \tag{1}$$

without imposing any restrictions on the growth of the functions $Q(s, t)$ for real $s = \sigma$.

It is natural to expect the correctness class to be relatively restrictive. It turns out that in general, it will contain only entire analytic functions the growth characteristics of which are suitably restricted.

We start with several propositions which generalize theorems from Section 4, Chapter III, Volume 2 to the case of a function $f(z)$ of arbitrary finite order.

We again restrict ourselves to the consideration of a single independent variable.

Let $f(z)$ be an entire analytic function of order $\leqslant \rho$ and type $\leqslant b^\rho$. This means that for any $\epsilon > 0$, the function satisfies the inequalities

$$|f(z)| \leqslant C_\epsilon \exp[(b + \epsilon)|z|]^\rho.$$

We denote by $\mathfrak{Z}_{\rho,b}$ the set of all entire functions of order ρ and type b^ρ. This set is obviously a linear system. Convergence in the set $\mathfrak{Z}_{\rho,b}$ is defined as follows: a sequence $f_\nu(z) \in \mathfrak{Z}_{\rho,b}$ converges by definition to zero for $\nu \to \infty$ if in any finite domain the sequence converges uniformly to zero and all functions $f_\nu(z)$ admit a common majorant of the form

$$C \exp(b|z|)^\rho.$$

Any function $f(z) \in \mathfrak{Z}_{\rho,b}$ defines a linear continuous functional

$$(f, \varphi) = \int \overline{f(x)}\, \varphi(x)\, dx$$

on the space K of infinitely differentiable functions of compact support. We wish to determine the Fourier transform $\tilde{f} = F(f)$ of the functional f. The functional $\tilde{f}$ is defined on the space Z, which is Fourier-dual to K. In the same manner as in Section 4, Chapter III, Volume 2, the result can be obtained by applying the Fourier operator F to every term of the Taylor expansion

$$f(z) = \sum a_\nu z^\nu, \tag{2}$$

It results that for every $\psi(s) \in Z$

$$(F(f), \psi) = \left(\sum a_\nu F(z^\nu), \psi\right) = \left(\sum a_\nu \left(-i\frac{\partial}{\partial\sigma}\right)^\nu \delta(\sigma), \psi(\sigma)\right)$$
$$= \sum (-i) a_\nu \frac{\partial^\nu \varphi(0)}{\partial \sigma^\nu}. \tag{3}$$

The series converges for any $\psi(s) \in Z$. One can even show that *the series converges for any entire analytic function of order* $\rho'\{(1/\rho) + (1/\rho') = 1\}$ *and type* $\leqslant (1/b_1)^{\rho'}$ *(b_1 will be defined below)*. Indeed, for any such function

$$g(s) = \sum \frac{s^\nu g^{(\nu)}(0)}{\nu!} \tag{4}$$

the values of the derivatives at the origin satisfy inequalities of the form

$$|g^{(\nu)}(0)| \leqslant C\nu! \frac{1}{b_1^{\ \nu}} \left(\frac{e\rho'}{\nu}\right)^{\nu/\rho'}, \tag{5}$$

whereas the Taylor series coefficients of the function $f(s)$ satisfy

$$|a_\nu| \leqslant AB^\nu \left(\frac{e\rho}{\nu}\right)^{\nu/\rho}. \tag{6}$$

Thus the series (3) for $g(s)$ has the majorant

$$AC \sum \left(\frac{b}{b_1}\right)^\nu \left(\frac{e\rho}{\nu}\right)^{\nu/\rho} \left(\frac{e\rho'}{\nu}\right)^{\nu/\rho'} \nu! = AC \sum \frac{e^\nu \nu!}{\nu^\nu} \left(\frac{b}{b_1}\right)^\nu (\rho^{1/\rho} \rho'^{1/\rho'})^\nu.$$

This latter series converges for

$$b_1 > b\rho^{1/\rho}\rho'^{1/\rho'},$$

whence we obtain an estimate for b_1. Consequently $F(f)$ is a functional on the space Z_{ρ_1, b_1}.

The continuity of this functional on Z_{ρ_1, b_1} follows from the fact that the constant C in inequality (5) depends on ν and vanishes as $\nu \to \infty$, hence $(F(f), \psi_\nu) \to 0$, whenever the sequence $\psi_\nu \in Z_{\rho_1, b_1}$ converges to zero as defined above.

Let $f_\nu(z) \in Z_{\rho, b}$ be a sequence which converges to zero in the topology of the space $Z_{\rho, b}$. We assert that the sequence of functionals $F(f_\nu)$ converges to zero in the topology of the dual space Z'_{ρ_1, b_1}, i.e., for any function $\psi(s) \in Z_{\rho_1, b_1}$

$$\lim_{\nu \to \infty} (F(f_\nu), \psi) = 0.$$

Under these assumptions, the constant A in (6) depends on ν, and for $\nu \to \infty$ it vanishes, which implies $(F(f_\nu), \psi) \to 0$, as required.

It can be easily seen that

$$(F(zf(z)), \psi(s)) = (F(f), i\psi'(s)),$$

Hence, for an arbitrary polynomial P

$$(F(P(z)f(z)), \psi(s)) = \left(F(f), P\left(i\frac{\partial}{\partial s}\right)\psi(s)\right). \tag{7}$$

Theorem 3. *If the resolvent matrix $Q(s, t)$ is of order $\leqslant p_0(>1)$ and type $\leqslant b^{p_0}$, then for sufficiently small t, there exists a solution of the Cauchy problem for the system (1) for any initial function $u(x)$ which can be continued into the complex $z = x + iy$ plane in such a manner as to yield an entire analytic function of order $\leqslant p_0'$ and type $\leqslant (1/b_1)^{p_0'}$.*

Proof. The inverse Fourier transform $G(x, t)$ of $Q(s, t)$ is a functional which can be extended to the space of entire functions of order $\leqslant p_0'$ and type $\leqslant (1/b_1)^{p_0'}$. In particular, the expression

$$(G(\xi, t), u(x - \xi))$$

is meaningful. We show that this function of x and t is the classical solution of the problem.

For $t = 0$ we have

$$(G(\xi, 0), u(x - \xi)) = (\delta(\xi), \quad u(x - \xi)) = u(x).$$

Differentiation with respect to t of $Q(s, t)$ is a continuous operation in $\mathfrak{Z}_{p,b}$. Hence the differentiation of the functional $G(x, t)$ with respect to t is continuous in the space $\mathfrak{Z}_{p_1,b_1}$, so that

$$\frac{\partial}{\partial t}(G(\xi, t), u(x - \xi)) = \left(\frac{\partial G(\xi, t)}{\partial t}, u(x - \xi)\right).$$

Taking into account the relation (7), we have further

$$\begin{aligned}\left(\frac{\partial G(\xi, t)}{\partial t}, u(x - \xi)\right) &= \left(F^{-1}\left(\frac{\partial}{\partial t}Q(s, t)\right), u(x - \xi)\right)\\ &= (F^{-1}(P(s)Q(s, t)), u(x - \xi))\\ &= \left(P\left(i\frac{\partial}{\partial \xi}\right)F^{-1}(Q(s, t)), u(x - \xi)\right)\\ &= \left(P\left(i\frac{\partial}{\partial \xi}\right)G(\xi, t), u(x - \xi)\right)\\ &= \left(G(\xi, t), P\left(-i\frac{\partial}{\partial \xi}\right)u(x - \xi)\right)\\ &= P\left(i\frac{\partial}{\partial x}\right)(G(\xi, t), u(x - \xi)).\end{aligned}$$

Consequently

$$u(x, t) = (G(\xi, t), u(x - \xi))$$

is indeed the solution of the Cauchy problem, we were looking for.

The following example shows that it is in general impossible to weaken the assumption of Theorem 3.

We consider the heat equation

$$\frac{\partial u}{\partial t} = \frac{\partial^2 u}{\partial x^2}$$

with an initial function $u_0(x)$ of compact support, which vanishes identically outside the interval $(-a, a)$. The solution of the problem

$$u(x, t) = \frac{1}{(2\pi t)^{1/2}} \int_{-a}^{a} \exp[-(x - \xi)^2/4t]\, u_0(\xi)\, d\xi \tag{8}$$

is obviously an entire analytic function of x, of order 2 and type arbitrarily close to $1/(4t)$ for any $t > 0$ and sufficiently small a. For definiteness, take $t = 1$. Consider the "inverse" heat equation

$$\frac{\partial v}{\partial t} = -\frac{\partial^2 v}{\partial x^2} \tag{9}$$

with the initial condition $v(x, 0) = u(x, 1)$. Obviously for $0 \leqslant t \leqslant 1$ the solution of this problem is $v(x, t) = u(x, 1 - t)$. For $t \to 1$, $v(x, t) \to u_0(x)$. For $t \to 1$ *the solution of Eq.* (9) *will no longer be a function.* Assuming the contrary leads to the following contradiction: choose a $\tau_0 > 1$ for which $v(x, \tau_0)$ is an ordinary function and take this function as initial function in the original problem. Then $u_0(x)$ would be a solution of the original equation for $t = 1 - \tau_0$ and would consequently be an analytic function. But this contradicts the assumption that $u_0(x)$ is of compact support. Consequently the Cauchy problem for Eq. (9) with initial function $u(x, 1)$ cannot admit an ordinary function as solution for $t > 1$.

CHAPTER IV

GENERALIZED EIGENFUNCTION EXPANSIONS

1. Introduction

The problem of eigenfunction expansions (or more generally, of expansions in eigenvectors of some operator) occurs in the most varied domains of mathematics. The problem of reducing a surface of second degree to its principal axes, or, what amounts to the same, the reduction of a quadratic form to its canonical form by means of an orthogonal transformation, is one of the simplest examples of such a problem.

Already at this simplest level, well known from linear algebra, to a quadratic form $A(x, x)$ and a given metric one associates a symmetric linear operator A according to the prescription

$$A(x, x) = (Ax, x).$$

The unit vectors along the principal axes of the surface $A(x, x) = 1$ are the (normalized) *eigenvectors* e_i of the operator A, i.e., they are n orthonormal vectors e_i such that the action of A reduces to multiplication with the *eigenvalues* λ_j of A:

$$Ae_j = \lambda_j e_j \qquad (j = 1, 2, \ldots, n).$$

Let $f = \sum_{j=1}^{n} f_j e_j$ be the expansion of the vector f in terms of the basis-vectors e_j (which, as assumed are normalized, i.e., $(e_i , e_j) = \delta_{ij}$). Then the action of A on the vector f is

$$Af = A\left(\sum_{j=1}^{n} f_j e_j \right) = \sum_{j=1}^{n} f_j \lambda_j e_j \,, \tag{1}$$

and the value of the quadratic form $A(f,f) = (Af,f)$ of f is given by

$$A(f,f) = \sum_{j=1}^{n} f_j \lambda_j^2. \tag{2}$$

One of the fundamental theorems of linear algebra is the theorem on the existence of a complete orthonormal basis formed out of eigenvectors for any symmetric linear operator A in n-dimensional Euclidean space.

The closest infinite-dimensional analog of an n-dimensional quadratic form is an integral form of the type

$$\int_a^b \int_a^b K(x, s) f(x) f(s)\, dx\, ds,$$

which occurs, for example, in the mechanics of systems with an infinite number of degrees of freedom (example: the energy for an inhomogeneous string). The reduction of such expressions to canonical forms is related to the problem of finding a complete system of (normalized) eigenfunctions of the symmetric integral operator

$$A\varphi(x) = \int_a^b K(x, s)\varphi(s)\, ds,$$

where $K(x, s) = K(s, x)$. Such a construction has been carried out by D. Hilbert in 1906.

For any such integral operator with the kernel $K(x, s)$ square-integrable in the rectangular region $a \leqslant s, x \leqslant b$, there exists a complete orthonormal set of eigenfunctions, i.e., functions $e_j(x) (j = 1, 2, \ldots)$ which satisfy the conditions

$$\int_a^b K(x, s) e_j(s)\, ds = \lambda_j e_j(x),$$

$$\int_a^b e_i(s) e_j(s)\, ds = \begin{cases} 1 & \text{for } i = j, \\ 0 & \text{for } i \neq j. \end{cases}$$

If $f = \sum_{j=1}^{\infty} f_j e_j(x)$ is the expansion of a function $f(x)$ in terms of the eigenfunctions $e_j(x)$, then the action of the operator A on f can be written

$$Af = A\left(\sum_{j=1}^{\infty} f_j e_j(x)\right) = \sum_{j=1}^{\infty} f_j \lambda_j e_j(x), \tag{3}$$

whereas the quadratic form (Af, f) becomes

$$(Af, f) = \int_a^b K(x, s) f(x) f(s)\, ds = \sum_{j=1}^{\infty} f_j \lambda_j^2. \tag{4}$$

As before, the numbers λ_j are called the *eigenvalues* (or characteristic values) of the operator A.

Subsequently, F. Riesz has given an abstract formulation of Hilbert's theorem which has also delineated the class of operators for which the theorem is valid. It turned out that the method is applicable to a relatively restricted class of self-adjoint operators in Hilbert space—the so-called *completely continuous*, or *compact operators*. A compact operator has the property of mapping any bounded set into a compact set; one can also characterize these operators as limits (in the norm-convergence of operators) of operators which map the whole Hilbert space into a finite dimensional space (operators of finite rank).

The method of separation of variables for linear partial differential equations also leads eigenvalue problems, this time however, for unbounded differential operators. In advantageous situations (finite domains, regular coefficients and suitable boundary conditions) one can reduce such problems by means of the Green's function to a symmetric integral equation. In more general problems, such a reduction becomes impossible and the problem of eigenfunction expansions must be studied separately. This kind of problem appears in the fundamental equations of quantum mechanics. Therefore, it became necessary to construct complete systems of eigenfunctions for general self-adjoint operators.

It is well known that even the simplest self-adjoint operators which are not compact in the Hilbert space of functions—like multiplication by x or differentiation (id/dx)—do not admit eigenvectors (i.e., eigenfunctions belonging to the Hilbert space).

It is useful to clarify the reasons why such operators do not admit eigenfunctions.

Let us consider as an example the operator of multiplication by x in the space $L_2(a, b)$ of square integrable functions on the interval $[a, b]$. Let us assume that for some function $y(x)$ we have

$$xy(x) = \lambda y(x). \tag{5}$$

This implies that $(x - \lambda)\, y(x) = 0$, i.e., that $y(x)$ vanishes except for $x = \lambda$. But there is no element of the space $L_2(a, b)$ except the null-function, which has this property.

As another example we consider the selfadjoint operator id/dx defined in the space $L_2(-\infty, \infty)$ (defined only for differentiable elements of this space, which form a dense subspace). An eigenfunction of this operator should satisfy the equation

$$i\frac{dy}{dx} = \lambda y. \tag{6}$$

But the solutions of this equation

$$y = Ce^{i\lambda x}$$

are not square-integrable for infinite intervals and hence do not belong to $L_2(-\infty, \infty)$.

Not wishing to go beyond the framework of a given Hilbert space H, mathematicians started to look for other, albeit weaker, formulations of the theorem on eigenfunction expansions. The first such formulation was essentially indicated by D. Hilbert in 1911 for the case of bounded self-adjoint operators A. (The multiplication by x in $L_2(a, b)$ is in particular such a bounded operator.)

Let A denote a bounded self-adjoint operator defined on the (separable) Hilbert space H:

$$(Ax, y) = (x, Ay)$$

for all x, y in H. It can be proved[1] that for each interval $\Delta = (\alpha, \beta)$ on the real axis there exists a maximal invariant subspace $H_\Delta \subset H$, for which the quadratic form (Ax, x) has values in the interval (α, β) for $\| x \| = 1$. This implies that that the restriction of the operator A to the subspace H_Δ realizes a mapping of that subspace into itself which differs in norm from multiplication by the number α by not more than $\beta - \alpha$. We denote the projection operator of the subspace H_Δ by $E(\Delta)$; $E(\Delta_{-\infty}^{\lambda})$, with $\Delta_{-\infty}^{\lambda}$ denoting the interval $(-\infty, \lambda)$ will be abbreviated as E_λ. The totality of projectors $E(\Delta)$ is called the spectral family of the operator A. If A is a compact (completely continuous) operator, for example the integral operator considered above, the spectral family $E(\Delta)$ consists of the projectors on the subspaces spanned by the eigenvectors for which the eigenvalues are situated in the interval Δ. In the general case, one can decompose the interval $[-\| A \|, \| A \|]$ of the real axis, for any $\epsilon > 0$, into a sum of intervals Δ_j ($j = 1, 2, \ldots, N$; $N \leqslant 2\| A \|/\epsilon$) of length smaller than ϵ such that the space H is decomposed into the orthogonal sum of subspaces H_Δ. In each of these subspaces, the operator A realizes a mapping which differs in norm by less than ϵ from multiplication with λ_j. Consequently the operator A differs (in norm) by less than ϵ from the operator

$$\sum_{j=1}^{n} \lambda_j E(\Delta_j).$$

throughout the whole space H.

[1] *Translator's note:* Cf. also for details Volume 4, pp. 127 ff. Most of this chapter should be read in parallel with Chapter I of Volume 4. Cf. also the book by Riesz and Nagy quoted in footnote 3, p. 169.

For $\epsilon \rightarrow 0$ this sum converges to a limit which can be denoted[2] as $\int_{-\infty}^{\infty} \lambda \, dE_\lambda$:

$$A = \int_{-\infty}^{\infty} \lambda \, dE_\lambda \,. \tag{7}$$

Actually, Eq. (7) is to be understood as

$$(Af, g) = \int \lambda d\,(E_\lambda f, g) \qquad \text{for all} \quad f, g \in H. \tag{8}$$

This is the "spectral resolution" of the operator A, which replaces the eigenvector expansion valid in the case of finite-dimensional or compact operators.

For $f = g$, (8) yields the spectral resolution of the quadratic form (Af, f):

$$(Af, f) = \int \lambda d\,(E_\lambda f, f) \tag{9}$$

as a generalization of Eqs. (2) and (4).

Later, von Neumann extended Hilbert's results to the most important case of an unbounded self-adjoint operator A which is defined on a dense subset H_A of the Hilbert space H. Let us recall the definition of an unbounded self-adjoint operator. In general, a bilinear form (Af, g) is not, for fixed g, a linear continuous functional of f. If for some g, this bilinear form is such a linear continuous functional, then there exists a $g_1 \in H$ (which is uniquely determined), such that $(Af, g) = (f, g_1)$. Thus, for the set of such g, a linear operator mapping g into g_1 is defined. This operator is called the *adjoint* of A and is denoted by A^*. Thus

$$(Af, g) = (f, A^* g).$$

If the operator A^* is defined on the domain of the operator A (if A is unbounded it is defined on a subspace of H, the domain of A) and on this domain coincides with A, then A is said to be *symmetric*.[3] If in addition the domains of the operators A and A^* coincide, the operator A is said to be *self-adjoint*. For self-adjoint operators, one can construct invariant subspaces H_Δ with the properties indicated above. But in this

[2] In fact the integration is over the interval $[-\|A\|, \|A\|]$.

[3] F. Riesz and B. Szökefalvy-Nagy, "Leçons d'analyse fonctionnelle." Akadémiai Kiadó, Budapest, 1952 (authors refer to Russian Ed. 1954). Engl. Transl. "Functional Analysis." Ungar, New York, 1955.

case, the interval $[-\| A \|, \| A \|]$ is replaced by the whole real axis, since an unbounded operator has no finite norm.

In the case of unbounded operators the distinction between symmetric and self-adjoint operators is essential. Thus, the operator id/dx, defined on the interval a, b for functions $y(x)$ which vanish at the points a and b (and are square integrable together with their derivative) is a symmetric operator, but is not self-adjoint. With these boundary conditions this operator does not admit eigenfunctions. If one replaces the boundary condition by the more general periodicity condition $y(a) = y(b)$, the operator will be self-adjoint (on the set we described above) and it admits a complete set of eigenfunctions.

It is very important to know under what conditions a given symmetric operator which is not self-adjoint can be extended into a self-adjoint operator (as shown by the preceding example). Such operators are called *essentially self-adjoint.*

We give here two sufficient conditions, each of which guarantees the existence of a self-adjoint extension of a given symmetric operator:[3]

(i) The operator A is *real*, i.e., its domain contains with each function f its complex conjugate $\bar{f}$, and $A\bar{f} = \overline{Af}$ (i.e., A maps real functions into real functions and complex conjugate pairs into complex conjugate pairs).

(ii) The operator A is *semi-bounded*, i.e., for all f in the domain of A we have the inequality

$$(Af, f) \geqslant \alpha(f, f)$$

(lower semi-boundedness). The largest admissible value of α is the infimum (inf) of the operator A. A semi-bounded symmetric operator admits a semi-bounded self-adjoint extension with the same inf.

But all these very substantial results represent only a surrogate for the spectral resolution, due to the fact that one imposes the condition of not going outside the initial Hilbert space framework.

Indeed, as is clear from the preceding examples of multiplication by x and differentiation (for functions defined on the whole real axis), it is in general impossible to obtain genuine eigenfunctions, without going beyond the limits of Hilbert space. The application to differential operators, for instance, of the general results cited above, notwithstanding their apparently definitive abstract formulations, leads to a series of difficulties.

On the other hand, there exists another possibility. It is suggested by the same examples of multiplication by x and differentiation. In both these examples eigenfunctions actually exist, but outside the limits of the original Hilbert space. For the operator of multiplication by x such eigenfunctions are the delta functions $\delta(x - \lambda)$ (a generalized function); for the operator of differentiation id/dx, the eigenfunctions are the nonsquare-integrable $e^{i\lambda x}$. Both these "eigenfunctions" can be obtained as linear functionals on a test function space which is a (dense) subspace of the initial Hilbert space. In some cases this space is the space K of infinitely differentiable functions of compact support.[4]

One can indicate a rather elementary procedure which leads from the spectral function E_λ to the required generalized eigenfunctions.

We have indicated that in the subspace corresponding to the projector $E(\Delta)$, $\Delta = (\alpha, \beta)$ the quadratic form (Ax, x) is for $\| x \| < 1$ bounded by

$$\alpha \leqslant (Ax, x) \leqslant \beta.$$

The operator A leaves the subspace H_Δ invariant. We define the operator $A - \lambda E$ which also maps the subspace H_Δ into itself. Since the norm of a symmetric operator is the least upper bound of the absolute values of the associated quadratic form, we have

$$\begin{aligned} \| A - \alpha E \| &= \sup_{(x,x)=1} ((A - \alpha E)x, x) \\ &= \sup_{(x,x)=1} ((Ax, x) - \alpha(x, x)) = \beta - \alpha. \end{aligned}$$

Therefore, for any normalized vector $x \in H_\Delta$

$$\|(A - \alpha E)x \| \leqslant \beta - \alpha.$$

Hence

$$Ax = \alpha x + \epsilon, \qquad \| \epsilon \| \leqslant \beta - \alpha.$$

We see that every vector in the space H_Δ is "almost" an eigenvector. The corresponding eigenvalue is α and the "error" is not larger than $\beta - \alpha$. The shorter the interval (α, β), the less does the vector Ax, $x \in H_\Delta$, differ from αx. It is clear that any genuine eigenvector corresponding to the eigenvalue α is situated in the intersection of all H_Δ as the interval Δ contracts to the point α.

But there may not be an element of the Hilbert space situated in this intersection—i.e., there may be no genuine eigenvector belonging to the eigenvalue α.

We may nevertheless make use of this construction in order to obtain an eigenvector in a wider space. With a fixed vector e, one can form the

[4] *Translator's note:* This is an example of a "rigged Hilbert space"—or Gel'fand triplet of spaces, which are discussed in detail in Volume 4.

vector $E(\Delta)\, e = (E_\beta - E_\alpha)\, e$, which belongs to the subspace H_Δ. We normalize this vector by dividing it with an appropriate number $\sigma(\Delta)$, and then let β converge to α.

The operation which was carried out with the vectors $e_\lambda = E_\lambda e$ is equivalent to a differentiation of this vector with respect to the parameter λ (for $\lambda = \alpha$) with a given measure σ_λ. It is clear that one must first investigate under what condition such a differentiation is legitimate. Therefore, we begin our discussion with the problem of differentiation of abstract functions depending on a parameter λ. It turns out that such an operation is possible, not in the framework of normed or Hilbert spaces, but in spaces of generalized functions. Thus spaces of generalized functions (distribution spaces) turn out to be the most convenient framework for spectral resolutions of differential operators.

We shall show in the following that the generalized eigenfunctions (or eigendistributions, as they are sometimes called) $\chi_\lambda = dE_\lambda\, e/d\sigma_\lambda$, obtained in this manner, form a complete set and we shall analyze the nature of these functions. In many cases, as for example for Sturm-Liouville problems, these will be ordinary functions. In this way, one can obtain the well-known results on spectral resolutions obtained by H. Weyl, F. Browder, L. Gårding, K. Kodaira, M. G. Kreĭn and F. Mautner. In addition, we obtain new spectral resolutions. Finally, we shall investigate the asymptotic behavior of the eigenfunctions. For nonelleptic differential operators the eigenfunctions are in general generalized functions.

2. Differentiation of Functionals of Strongly Bounded Variation

2.1. Functionals over a Banach Space

We first consider a Banach space Φ and its dual Φ' formed by the linear continuous functionals over Φ. Φ' is also complete, i.e., a Banach space.

A linear functional f_λ which depends on the parameter λ ($a \leqslant \lambda \leqslant b$), is said to be of *strongly bounded variation*, if for any partition of the interval $a \leqslant \lambda \leqslant b$

$$a = \lambda_0 < \lambda_1 < \cdots < \lambda_n = b,$$

the norms satisfy the inequality

$$\sum \| f_{\lambda_{j+1}} - f_{\lambda_j} \| < C, \tag{1}$$

where C is a constant.

Applying the functional f_λ to a fixed element of the space Φ, one obtains a numerical function (f_λ, φ) of λ which will be of bounded variation in the usual sense, since

$$\sum |(f_{\lambda_{j+1}}, \varphi) - (f_{\lambda_j}, \varphi)| = \sum |(f_{\lambda_{j+1}} - f_{\lambda_j}, \varphi)| \leqslant \sum \| f_{\lambda_{j+1}} - f_{\lambda_j} \| \cdot \| \varphi \|. \quad (2)$$

A function of bounded variation admits a derivative almost everywhere.[5] In particular, the numerical function (f_λ, φ) is differentiable with respect to λ, possibly with the exception of a set of measure zero. This set of measure zero will in general depend on the vector φ. The question arises, whether one can take the same set of measure zero for all φ. If the answer to this question were affirmative, one could assert the existence for almost all λ of a functional g_λ, which is the weak derivative of f_λ with respect to λ:

$$g_\lambda = \lim_{\Delta\lambda \to 0} \frac{f_{\lambda+\Delta\lambda} - f_\lambda}{\Delta\lambda} = \frac{df_\lambda}{d\lambda}. \quad (3)$$

It turns out that this is indeed so if the space Φ is *separable*, i.e., admits a countable everywhere dense set $\{\varphi_\nu\}$. The resulting theorem is valid not only for Lebesgue measure but for any σ-additive measure on the interval $a \leqslant \lambda \leqslant b$.

Theorem 1. *If the Banach space Φ is separable, then any linear continuous functional f_λ, defined on Φ for $a \leqslant \lambda \leqslant b$, which is of strongly bounded variation with respect to λ, is almost everywhere weakly differentiable with respect to any nonnegative σ-additive measure μ defined on the Borel sets of the closed interval $[a, b]$.*

Proof. We denote by $\mu(\Delta_\alpha^\beta)$ the value of the measure for the semi-closed interval $\alpha \leqslant \lambda < \beta$. We show first that the function $\| f_\lambda \|$ is almost everywhere Lipschitzian:

$$\| f_{\lambda+h} - f_\lambda \| \leqslant c_\lambda \mu(\Delta_\lambda^{\lambda+h}). \quad (4)$$

If the inequality (4) were not true, there would exist a set P of positive measure $\mu(P) = \gamma > 0$ such that for any N and any $\lambda \in P$ there is a sequence $\lambda_N^{(m)} \to \lambda$ for which

$$\| f_{\lambda_N^{(m)}} - f_\lambda \| > N \mu(\Delta_\lambda^{\lambda_N^{(m)}}).$$

[5] Cf. e.g., I. P. Natanson, "Teoriya funktsii veshchestvennoi peremennoi" (Theory of Functions of a Real Variable) p. 193. Gostekhizdat, Moscow, 1950 (or any English text on real variables).

According to Vitali's theorem,[6] one can select from the system of intervals $\Delta_{\lambda}^{\lambda(m)}{}_N$ covering P, a finite number of disjoint intervals which together cover a set Q, with $\mu(Q) > \gamma/2$. Let (λ_j, λ_j') denote these intervals ($j = 1, 2, \ldots$), then

$$\sum \| f_{\lambda'_j} - f_{\lambda_j} \| > N \sum \mu(\Delta_{\lambda_j}^{\lambda'_j}) > \tfrac{1}{2} N\gamma;$$

but this contradicts the assumption of strongly bounded variation for f_λ.

For any $\varphi \in \Phi$ the numerical function (f_λ, φ) is of bounded variation. Indeed

$$\sum |(f_{\lambda_{j+1}}, \varphi) - (f_{\lambda_j}, \varphi)| \leqslant \| \varphi \| \sum \| f_{\lambda_{j+1}} - f_{\lambda_j} \| \leqslant C \| \varphi \|,$$

and consequently (f_λ, φ) admits almost everywhere a derivative with respect to the measure μ. This is in particular true for the elements φ_ν of a countable dense set.

Consider a set Q of total μ-measure (i.e., such that $\mu(Q \setminus [a, b]) = 0$) on which the function $\| f_\lambda \|$ satisfies the Lipschitz condition and all functions (f_λ, φ_ν) ($\nu = 1, 2, \ldots$) admit derivatives. We assert that all functions (f_λ, φ) have derivatives on the set Q. Indeed, for any $\lambda \in Q$ we have, on the one hand

$$\frac{\| f_{\lambda+h} - f_\lambda \|}{\mu(\Delta_\lambda^{\lambda+h})} < C_\lambda ;$$

on the other hand, for arbitrary h there exists a limit for the ratio

$$\frac{(f_{\lambda+h}, \varphi_\nu) - (f_\lambda, \varphi_\nu)}{\mu(\Delta_\lambda^{\lambda+h})}$$

Thus the system of functionals

$$\frac{f_{\lambda+h} - f_\lambda}{\mu(\Delta_\lambda^{\lambda+h})}$$

is bounded in norm for fixed $\lambda \in Q$ and converges on the elements of a countable dense set $\{\varphi_\nu\}$.

But a sequence of continuous linear functionals which are bounded in norm and converges on a countable dense set in the separable Banach space Φ is convergent for any element of the space Φ.

Consequently the functional f_ν, for $\lambda \in Q$, admits a weak derivative χ_λ with respect to the measure μ as required.

[6] In the book by Natanson quoted in the preceding footnote the proof is given for Lebesgue measure. It is easy to generalize the proof to the case of a general nonnegative σ-additive measure, which we require here.

2.2. Functionals over Countably-Normed Spaces

We recall from Sections 3 and 4, Chapter I, Volume 2, the fundamental facts about the structure of (complete) countably-normed spaces Φ. Such a space is the intersection of a sequence of complete normed (Banach-) spaces

$$\Phi_1 \supset \Phi_2 \supset \cdots \supset \Phi_\nu \supset \cdots \supset \Phi$$

with increasing norms

$$\| \varphi \|_1 \leqslant \| \varphi \|_2 \leqslant \cdots \leqslant \| \varphi \|_\nu \leqslant \cdots,$$

which are coordinated with each other in the sense that if a sequence $\{\varphi_\nu\}$ is a Cauchy sequence in the norm $\| \, \|_{k+m}$ and converges to zero in the norm $\| \, \|_k$, then it also converges to zero in the norm $\| \, \|_{k+m}$. The space Φ_p is to be understood as the completion of the space Φ in the norm $\| \, \|_p$.

Any continuous linear functional over the Banach space Φ_p is at the same time a linear continuous functional over the space Φ. It has been proved in Chapter I of Volume 2, that the converse is also true: each continuous linear functional over the space Φ can be extended to a linear continuous functional over some space Φ_p and is therefore an element of the dual Φ_p'. The set of all linear functionals over Φ which forms the dual space Φ' is thus the union of all dual spaces Φ_p':

$$\Phi_1' \subset \Phi_2' \subset \cdots \subset \Phi_p' \subset \cdots \subset \Phi' = \bigcup_{p=1}^{\infty} \Phi_p'.$$

The smallest number p for which a functional $f \in \Phi'$ belongs to the space Φ_p' is called the *order* of the functional. In each of the spaces Φ_p', Φ_{p+1}', ... the functional f has a norm, satisfying the decreasing sequence of inequalities

$$\| f \|_p \geqslant \| f \|_{p+1} \geqslant \cdots .$$

Formally one can add to this sequence the first terms $\| f \|_1, \ldots, \| f \|_{p-1}$, considering them equal to infinity.

By definition, a functional f, defined on a countably-normed space Φ is of *strongly bounded variation*, if there is a number p such that the functional f belongs to the space Φ_p' and is of strongly bounded variation with respect to the norm of that space, i.e.,

$$\sum \| f_{\lambda_{j+1}} - f_{\lambda_j} \|_p < C.$$

Making use of Theorem 1 of the preceding section we obtain:

Theorem 2. *A linear continuous functional f_λ of strongly bounded variation (with respect to λ), defined on a separable countably-normed space Φ is almost everywhere weakly differentiable with respect to a positive σ-additive measure μ. Its weak derivative*

$$\gamma_\lambda = \frac{df_\lambda}{d\mu}$$

is a functional belonging to that space Φ_p', with respect to the norm of which the functional f_λ is of strongly bounded variation.

3. Differentiation of Functionals of Weakly Bounded Variation

3.1. General Considerations

We consider the continuous linear functional f_λ, defined on the linear topological space Φ and depending in the parameter λ, $a \leqslant \lambda \leqslant b$. To each element φ of the space this functional associates a numerical function of λ equal to (f_λ, φ). The functional f_λ is said to be of *weakly bounded* variation in $[a, b]$ if the numerical function (f_λ, φ) is of bounded variation for any $\varphi \in \Phi$, i.e., if for each partition of the interval $a \leqslant \lambda \leqslant b$ by means of the intermediate points $a = \lambda_0 < \lambda_1 < \cdots < \lambda_n = b$, the inequality

$$\sum |(f_{\lambda_{j+1}}, \varphi) - (f_{\lambda_j}, \varphi)| \equiv \sum |(f_{\lambda_{j+1}} - f_{\lambda_j}, \varphi)| \leqslant C_\varphi \tag{1}$$

holds with a constant C_φ depending only on the element φ.

Example 1. We consider one of the function spaces over the line $-\infty < x < \infty$ consisting of absolutely integrable functions, and the functional over this space defined by the step function

$$\theta(x - \lambda) = \begin{cases} 0 & \text{for } x < \lambda \\ 1 & \text{for } x > \lambda: \end{cases}$$

$$(\theta(x - \lambda), \varphi(x)) = \int_\lambda^\infty \varphi(x)\, dx.$$

We have

$$\sum |(\theta(x - \lambda_{j+1}) - \theta(x - \lambda_j), \varphi)| = \sum \left| \int_{\lambda_j}^{\lambda_{j+1}} \varphi(x)\, dx \right| \leqslant \int_{-\infty}^{\infty} |\varphi(x)|\, dx,$$

i.e., condition (1) is fulfilled.

Example 2. In a Hilbert space, we consider the spectral family E_λ and a fixed vector e. The linear functional $(E_\lambda e, \varphi)$, is also of weakly bounded variation.

Let $\{\lambda_j\}$ denote a partition of the λ axis into a finite number of intervals Δ_j and let $E(\Delta_j) = E_{\lambda_{j+1}} - E_{\lambda_j}$ $(j = 1, 2, \ldots)$ denote the corresponding projectors. Setting $e_j = E(\Delta_j)\, e$ and $\varphi_j = E(\Delta_j)\, \varphi$ we obtain

$$(E(\Delta_j)e, \varphi) = (E(\Delta_j)\, e, E(\Delta_j)\varphi) = (e_j\, , \varphi_j),$$

$$|(e_j, \varphi_j)| \leqslant \| e_j \| \cdot \| \varphi_j \| \leqslant \tfrac{1}{2}(\| e_j \|^2 + \| \varphi_j \|^2),$$

$$\begin{aligned} \sum |((E_{\lambda_{j+1}} - E_{\lambda_j})e, \varphi)| &= \sum |(E(\Delta_j)e, \varphi)| \\ &\leqslant \tfrac{1}{2} \sum (\| e_j \|^2 + \| \varphi_j \|^2) \\ &= \tfrac{1}{2}(\| e \|^2 + \| \varphi \|^2), \end{aligned}$$

i.e., the inequality (1) is satisfied.

One can give a different formulation of the definition of the concept of a functional of weakly bounded variation. It follows from inequality (1) that for arbitrarily chosen numbers ϵ_j $(| \epsilon_j | = 1)$,

$$| \sum \epsilon_j (f_{\lambda_{j+1}} - f_{\lambda_j}\, , \varphi)| \leqslant C_\varphi\, ,$$

or

$$| (\sum \epsilon_j (f_{\lambda_{j+1}} - f_{\lambda_j}), \varphi)| \leqslant C_\varphi\, .$$

The last inequality signifies that the set of all functionals of the form

$$\sum \epsilon_j (f_{\lambda_{j+1}} - f_{\lambda_j}) \qquad (| \epsilon_j | = 1) \tag{2}$$

is a bounded set in Φ' (i.e., is weakly bounded and consequently also, strongly bounded; cf. Section 5, Chapter I, Volume 2).

Conversely, if the set of all functionals of the form (2) forms a bounded set in Φ', this means that for an arbitrary choice of the numbers ϵ_j $(| \epsilon_j | = 1)$ and an arbitrary element φ the inequality

$$|(\sum \epsilon_j (f_{\lambda_{j+1}} - f_{\lambda_j}), \varphi)| \leqslant C_\varphi\, .$$

holds. If one chooses the ϵ_j so that

$$(\epsilon_j (f_{\lambda_{j+1}} - f_{\lambda_j}), \varphi) = |(f_{\lambda_{j+1}} - f_{\lambda_j}\, , \varphi)|,$$

we arrive at the condition (2).

We have thus proved that: *A functional f is of weakly bounded variation* *if and only* if all *functionals of the form* (2) *form a bounded set in the space* Φ'.

In the same manner as for functionals of strongly bounded variation, there arises the problem of differentiability of a functional of weakly bounded variation.

This question is important not only from the general point of view. The differentiation of the functional defined by the function $\theta(x - \lambda)$ (Example 1.) leads to a delta function, which is a generalized function. Similarly, the differentiation of the functional $(E_\lambda e, \varphi)$ could lead to generalized eigenvectors of the operator A, to which the spectral family E_λ belongs.

But these same examples show that the problem of differentiability of a linear functional of weakly bounded variation cannot in general be answered in the affirmative in the framework of Banach spaces (or more particularly, Hilbert space). Indeed, the function $\theta(x - \lambda)$ is an element of a Hilbert space (square integrable functions on an interval) but its derivative $\delta(x - \lambda)$ is no longer an element of this space.

The answer to the question posed above is affirmative if the functional is not only of weakly bounded variation, but also of strongly bounded variation. This is usually not the case for functionals defined on normed spaces. In particular it is not true for such functionals as $\theta(x - \lambda)$ or $E_\lambda e$.

The situation is radically different for functionals defined on countably normed spaces[7] where it is more often the case that a functional of weakly bounded variation is also of strongly bounded variation. This is true in particular for the so-called *nuclear spaces*.

A countably normed space will be called a *nuclear space* or *N-space* if each continuous linear functional f_λ which is of weakly bounded variation with respect to λ is also of strongly bounded variation.[8] *In each separable, and in particular, in each perfect N-space*[9] *any functional* f_λ *of weakly bounded variation admits a weak derivative, almost everywhere with respect to any nonnegative* σ*-additive measure.*

3.2. The Case of the Space $K\{M_p\}$

We show that a space $\Phi = K\{M_p\}$ (cf. Section 1, Chapter II, Volume 2) is an N-space, under certain conditions. A space $K\{M_p\}$

[7] *Translator's note:* Or countably-Hilbert spaces, cf. Volume 4.

[8] *Translator's note:* In Volume 4 a different definition of nuclearity is given for countably Hilbert spaces. Under certain assumptions about the functions $M_p(x)$, the nuclearity of the spaces $K(M_p)$ is also proved there.

[9] *Translator's note:* In Section 3, Chapter I, Volume 4, it will be proved that each nuclear space is perfect.

consists of all infinitely differentiable functions $\varphi(x)$ which are such that the products $M_p(x)| D^q\varphi(x)|$ are continuous and bounded in R.[10] Here $1 \leqslant M_0(x) \leqslant \cdots \leqslant M_p(x) \leqslant \cdots$ is a given sequence of functions. The functions $M_p(x)$ are either all infinite for a given x, or all are finite, for all p. Wherever these functions are finite they are assumed to be continuous. We asume that the conditions (P) and (N) are satisfied. The condition (P) consists in the existence for each p of a $p' > p$, such that

$$\lim_{|x|\to\infty} \frac{M_p(x)}{M_{p'}(x)} = 0;$$

if this condition is satisfied, the space $K\{M_p\}$ is perfect (Section 2, Chapter II, Volume 2), i.e., all its bounded sets are compact. The condition (N) requires that for each p, there exist a $p' > p$, such that

$$\frac{M_p(x)}{M_{p'}(x)} = m_{pp'}(x)$$

is an integrable function of x. In a space $K\{M_p\}$ satisfying condition (N) one can introduce two equivalent systems of norms

$$\| \varphi \|_p' = \sup_{x,\, |q| \leqslant p} M_p(x) \mid D^q\varphi(x)|, \tag{a}$$

$$\| \varphi \|_p' = \sup_{|q| \leqslant p} \int_{-\infty}^{\infty} M_p(x) \mid D^q\varphi(x)| \, dx. \tag{b}$$

Let Φ_p denote the completion of the space Φ with respect to the norm (a) and let Φ^p denote its completion with respect to the norm (b).

According to the general theory, the dual space Φ' can be viewed as the union of the dual spaces Φ_p' and at the same time as the union of the dual spaces $\Phi^{p\prime}$.

Let $f_\lambda \in \Phi'$ be a functional of weakly bounded variation for $a \leqslant \lambda \leqslant b$. As shown in Section 3.1, this means that all functionals of the form $\sum_{j=0}^{n-1} \epsilon_j(f_{\lambda_{j+1}} - f_{\lambda_j})$ form a bounded set in Φ' for arbitrarily chosen points λ_j and numbers $\epsilon_j(| \epsilon_j | = 1)$. But then all these functionals belong to a space $\Phi^{p\prime}$ for a certain p, and form a bounded set in $\Phi^{p\prime}$. At the same time, the functionals f_λ form themselves a bounded set, so that one can assume, without loss of generality, that the f_λ belong to $\Phi^{p\prime}$ and are bounded there with respect to their norm. The general form of a con-

[10] Here, $D^q = \dfrac{\partial^{|q|}}{\partial x_1^{q_1} \cdots \partial x_n^{q_n}}, \qquad | q | = q_1 + \cdots + q_n \leqslant p.$

tinuous linear functional on a space Φ^p was determined in Section 4.2, Chapter II, Volume 2. It was shown there that

$$(f_\lambda, \varphi) = \sum_{|q| \leqslant p} \int M_p(x) f_\lambda^q(x) D^q \varphi(x)\, dx, \tag{1}$$

where $f_\lambda^q(x)$ is for each q and λ a measurable and bounded function in the space R. The norm of the functional is defined as

$$\| f_\lambda \|_{\Phi^{p'}} = \sum_{|q| \leqslant p} \sup_x | f_\lambda^q(x)|,$$

where sets of measure zero are to be neglected in taking the sup. Furthermore

$$\Big(\sum_j \epsilon_j (f_{\lambda_{j+1}} - f_{\lambda_j}), \varphi\Big) = \sum_q \int M_p(x) \Big\{\sum_j \epsilon_j (f^q_{\lambda_{j+1}}(x) - f^q_{\lambda_j}(x)\Big\} D^q \varphi\, dx.$$

Since the norms of these functionals are bounded, we have

$$\sum_{|q| \leqslant p} \sup_x \Big| \sum_j \epsilon_j (f^q_{\lambda_{j+1}}(x) - f^q_{\lambda_j}(x)) \Big| \leqslant C.$$

For every fixed x the multipliers ϵ_j, $| \epsilon_j | = 1$, can be chosen arbitrarily, therefore

$$\sum_{|q| \leqslant p} \sup_x \sum_j | f^q_{\lambda_{j+1}}(x) - f^q_{\lambda_j}(x)| \leqslant C. \tag{2}$$

We estimate now the norm of the functional $f_{\lambda_{j+1}} - f_{\lambda_j}$ in the space $\Phi'_{p'}$, where p' is chosen according to condition (N), to agree with p. For $\varphi \in \Phi_{p'}$:

$$\begin{aligned} |(f_{\lambda_{j+1}} - f_{\lambda_j}, \varphi)| &\leqslant \sum_{|q| \leqslant p} \int M_p(x) \,| f^q_{\lambda_{j+1}}(x) - f^q_{\lambda_j}(x)| \cdot | D^q \varphi(x)|\, dx \\ &\leqslant \sum_{|q| \leqslant p} \int m_{pp'}(x) \,| f^q_{\lambda_{j+1}}(x) - f^q_{\lambda_j}(x)| \cdot M_{p'} \,| D^q \varphi(x)|\, dx \\ &\leqslant \sup_x M_{p'}(x) \,| D^q \varphi(x)| \cdot \sum_{|q| \leqslant p} \int m_{pp'} \,| f^q_{\lambda_{j+1}}(x) - f^q_{\lambda_j}(x)|\, dx, \end{aligned}$$

and consequently

$$\| f_{\lambda_{j+1}} - f_{\lambda_j} \|_{\Phi'_{p'}} \leqslant \sum_{|q| \leqslant p} \int m_{pp'}(x) \,| f^q_{\lambda_{j+1}}(x) - f^q_{\lambda_j}(x)|\, dx.$$

Making use of (2), we find

$$\sum_j \|f_{\lambda_{j+1}} - f_{\lambda_j}\|_{\Phi'_{p'}} \leqslant \sum_{|q|\leqslant p} \int m_{pp'}(x) \sum_j |f^q_{\lambda_{j+1}}(x) - f^q_{\lambda_j}(x)|\, dx$$

$$\leqslant C \int m_{pp'}(x)\, dx = C_{pp'},$$

whence

$$\sup \sum_j \|f_{\lambda_{j+1}} - f_{\lambda_j}\|_{\Phi'_{p'}} \leqslant C_{pp'}, \tag{3}$$

as required.

Remembering that a perfect space Φ is separable and making use of the theorem from Section 1 we conclude that *in a space* $\Phi = K\{M_p\}$ *which satisfies the conditions (N) and (P), any functional* f_λ $(a \leqslant \lambda \leqslant b)$ *of weakly bounded variation admits a derivative with respect to any non-negative σ-additive measure* $\mu(\lambda)$ *almost everywhere in* $[a, b]$; *this derivative is a functional* $df_\lambda / d\mu(\lambda)$ *belonging to the same space* Φ_p' *for which the expressions*

$$\sum_{j=0}^{n} \|f_{\lambda_{j+1}} - f_{\lambda_j}\|_{p'}.$$

are bounded.

We recall that the space $\Phi_{p'}$ consists of functions $\varphi(x)$ which are p' times continuously differentiable and have finite norm

$$\|\varphi\|_{p'} = \sup_{\substack{x \\ |q|\leqslant p'}} M_{p'}(x)\, |D^q\varphi(x)|.$$

There exists another definition of an N-space (nuclear space), which will not be used in this chapter, but which we give here, due to its simplicity. For details and applications the reader is referred to Section 3 and following, in Chapter I, Volume 4, in particular, the nuclearity of the space $K\{M_p\}$ is proved in Section 3.6, there.

A series of functionals $f_1 + f_2 + \cdots + f_k + \cdots$ is said to be *unconditionally convergent* if for any test function φ the series

$$\sum_{j=1}^{\infty} |(f_j, \varphi)|.$$

converges. The same series is said to be *absolutely convergent* if there exists a neighborhood of zero U in Φ such that the series

$$\sum_{j=1}^{\infty} \|f_j\|_U$$

converges, with the norm $\|f_j\|_U = \sup_{\varphi\in U} |(f_j, \varphi)|$.

One can define an N-space as a space for which *every unconditionally convergent series of functionals is absolutely convergent.* A proof of this statement can be found in Volume 4, p. 67 ff.

4. Existence and Completeness Theorems for the System of Eigenfunctionals

We show in this section that a spectral family E_λ defined on a given Hilbert space H can be differentiated with respect to the parameter λ in terms of a measure σ. The result of this differentiation $\chi_\lambda = dE_\lambda e/d\sigma(\lambda)$ is, for fixed λ, a generalized function and the set of all generalized eigenfunctions χ_λ obtained in this manner forms, in a sense to be defined below, a complete system.

4.1. General Remarks

We consider a perfect space Φ, i.e., a countably normed space in which every bounded set is relatively compact (i.e., has compact closure).

In addition to the original countably-normed topology, let there be an inner-product metric defined on Φ: to each pair of elements φ, ψ there is a complex number[11] (φ, ψ) (the inner product) with the usual properties:

$$(1) \qquad (\varphi_1 + \varphi_2, \psi) = (\varphi_1, \psi) + (\varphi_2, \psi);$$

$$(2) \qquad (\varphi, \psi) = \overline{(\psi, \varphi)};$$

$$(3) \qquad (\alpha\varphi, \psi) = \alpha(\varphi, \psi);$$

$$(4) \qquad (\varphi, \varphi) > 0 \quad \text{for} \quad \varphi \neq 0.$$

In addition, we assume that the following condition is satisfied

(5) If $\varphi_\nu \to \varphi$ in the (countably-normed) original topology of Φ then $(\varphi_\nu, \psi) \to (\varphi, \psi)$ for each ψ (continuity).

The inner product being a bilinear form, it associates to each element $\varphi \in \Phi$ a linear continuous functional f_φ according to the formula

$$(f_\varphi, \psi) = (\varphi, \psi).$$

[11] It should be clear from the context when (f, φ) denotes the value of the linear functional f on the element φ and when (φ, ψ) denotes an inner product on Φ.

The association $\varphi \leftrightarrow f_\varphi$ is obviously linear. To different elements of the space Φ it associates different elements of the space Φ': if $f_{\varphi_1} = f_{\varphi_2}, f_{\varphi_1 - \varphi_2} = 0$, consequently $(\varphi_1 - \varphi_2, \psi) = 0$ for all $\psi \in \Phi$; with $\psi = \varphi_1 - \varphi_2$ we have $(\varphi_1 - \varphi_2, \varphi_1 - \varphi_2) = 0$, whence, by (4), $\varphi_1 = \varphi_2$. Finally the association is *continuous*: if $\varphi_\nu \to \varphi_0$, then condition (5) implies $(\varphi_\nu, \psi) \to (\varphi_0, \psi)$ for any $\psi \in \Phi$ and consequently $f_{\varphi_\nu} \to f_{\varphi_0}$ weakly (and due to the fact that Φ is a perfect space, this implies strong convergence).

The inner product (φ, ψ) makes Φ into a pre-Hilbert space and its completion with respect to the metric defined by the inner product makes it into a Hilbert space H. It can be shown that the mapping $\varphi \to f_\varphi$ can be extended from the space Φ to H. Indeed, any element $h \in H$ defines a linear functional on Φ via the inner product (h, φ); this functional is continuous in the topology of H. We verify that it is also continuous in the topology of Φ. Let $\varphi_\nu \to h$ in the topology of H. Then $(\varphi_\nu, \psi) \to (h, \psi)$, i.e., the functional h is a weak limit of continuous linear functionals, and according to the theorem of Section 5.6, Chapter I, Volume 2, is itself a continuous linear functional. Thus, to each $h \in H$ one can associate a linear continuous functional $f_h \in \Phi'$, such that $(f_h, \varphi) = (h, \varphi)$ (here the left side brackets denote a functional, the right hand brackets denote an inner product—the meaning should be clear from the context). A reasoning similar to the one given above shows that the mapping $h \to f_h$ is continuous and one-to-one. Identifying the functionals f_h with the corresponding elements h, we obtain the inclusions[12]

$$\Phi \subset H \subset \Phi'.$$

Example 1. Let Φ denote one of the perfect spaces of square-integrable functions $\varphi(x)$ $(-\infty < x < \infty)$, for example the space $K\{M_p\}$, satisfying condition (P) (cf. Section 3.2) and define

$$(\varphi, \psi) = \int_{-\infty}^{\infty} \varphi(x)\overline{\psi(x)}\,dx. \tag{1}$$

In this case the Hilbert space is $L_2(-\infty, \infty)$. The injection $\varphi \to f_\varphi$ is the identity map.

[12] *Translator's note:* The careful reader will have noticed that if the three spaces involved in the Gelfand-triplet ("rigged Hilbert space") are considered as vector spaces over the field of complex numbers, the sesquilinearity of the inner product in H implies that one of the two inclusions should be understood "antilinearly" (i.e., functions should be replaced by their complex conjugates, when making the appropriate identifications); cf. Volume 4, p. 107. Note that in Volume 4 the *dual* of a space is called the *conjugate*!

Example 2. Let Φ be one of the similar spaces of functions of two variables $\varphi(x_1, x_2)$ and let

$$(\varphi, \psi) = \int\int \left(\frac{\partial \varphi}{\partial x_1} \frac{\overline{\partial \psi}}{\partial x_1} + \frac{\partial \varphi}{\partial x_2} \frac{\overline{\partial \psi}}{\partial x_2} \right) dx_1 \, dx_2 \, . \tag{2}$$

Taking into account the decrease of these functions and of their derivatives at infinity, one obtains, making use of the Green identity:

$$\int\int \left(\frac{\partial \varphi}{\partial x_1} \frac{\overline{\partial \psi}}{\partial x_1} + \frac{\partial \varphi}{\partial x_2} \frac{\overline{\partial \psi}}{\partial x_2} \right) dx_1 \, dx_2 = \int\int \bar{\psi} \left(\frac{\partial^2 \varphi}{\partial x_1{}^2} + \frac{\partial^2 \varphi}{\partial x_2{}^2} \right) dx_1 \, dx_2 \, .$$

Here H is the Hilbert space obtained by completion in the inner product (2) and the mapping $\varphi \to f_\varphi$ is obtained by applying the Laplace operator to the function φ.

Example 3. Let A be an arbitrary symmetric positive definite differential operator. We can define for an element φ of the test function space Φ

$$(\varphi, \psi) = \int \bar{\psi} \cdot A\varphi \, dx.$$

Here the mapping $\varphi \to f_\varphi$ is the application of the operator A to φ.

4.2. The Existence of Eigenfunctionals (Generalized Eigenfunctions)

As already indicated, the problem consists in establishing the existence of the derivative of a given spectral family E_λ with respect to a measure.

We define this measure in the following manner. Let e denote a normed vector of the space H and let $H(e)$ denote the subspace spanned by the vectors $e_\lambda = E_\lambda e$. The immediately following constructions will be in terms of the space $H(e)$. We define the function $\sigma(\lambda) = (E_\lambda e, e)$. On the basis of the known properties of the spectral resolution this is a monotone function of λ taking values from 0 to 1 as λ varies from $-\infty$ to $+\infty$. Such a function generates in a well-known manner a (Lebesgue)-Stieltjes measure on the axis $-\infty < \lambda < +\infty$: the measure of a set P is given by the integral $\sigma_\lambda(P) = \int_P d\sigma(\lambda)$.

We now formulate and prove the first of our fundamental theorems.

Theorem 1. *Let H be a Hilbert space obtained by completion of a (perfect) N-space Φ with respect to the inner product (φ, ψ) and let $E_\lambda = E(\Delta^\lambda_{-\infty})$ be a spectral family of projectors defined on H. Let further e*

be a fixed normed vector in H and define $\sigma(\lambda) = (E_\lambda e, e)$. Then, almost everywhere with respect to the measure $\sigma_\lambda = \sigma(\lambda)$ there exists the derivative

$$\frac{dE_\lambda e}{d\sigma_\lambda} = \chi_\lambda ,$$

and this derivative is a continuous linear functional defined on Φ. Its action on an element $\varphi \in \Phi$ is given by

$$(\chi_\lambda , \varphi) = \frac{d(E_\lambda e, \varphi)}{d\sigma_\lambda} . \tag{1}$$

Proof. As an abstract function with values in Φ' the function $E_\lambda e = e_\lambda$ is of weakly bounded variation. Indeed, for any $\varphi \in \Phi$ the bilinear form $(e_\lambda , \varphi) = (E_\lambda e, \varphi)$ of the vectors e_λ and φ can be represented as a linear combination of four quadratic forms, each a monotone function of λ:

$$\begin{aligned}(E_\lambda e, \varphi) = \tfrac{1}{4}\{(E_\lambda(e + \varphi), e + \varphi) + i(E_\lambda(e + i\varphi), e + i\varphi) \\ - (E_\lambda(e - \varphi), e - \varphi) - i(E_\lambda(e - i\varphi), e - i\varphi)\}.\end{aligned}$$

Since Φ is a nuclear space, it follows that $E_\lambda e$ is a functional of weakly bounded variation.

But then according to Section 2, there exist (for almost all λ with respect to the measure σ_λ) the functionals

$$\chi_\lambda = \frac{dE_\lambda e}{d\sigma_\lambda} ,$$

which operate according to the formula

$$(\chi_\lambda , \varphi) = \frac{d(E_\lambda e, \varphi)}{d\sigma_\lambda} .$$

Thus Theorem (1) is proved.

4.3. The Completeness of the System of Eigenfunctionals

In this section we analyze the problem of orthogonality and completeness of the system of generalized eigenfunctions (eigenfunctionals) χ_λ.

Naturally, here it is meaningless to talk about ordinary orthogonality, since inner products of the form $(\chi_\lambda , \chi_\mu)$ are not defined. This is also true in classical analysis, e.g., in the theory of Fourier integrals, where one cannot talk about orthogonality of the functions $e^{i\lambda x}$ on the whole axis; a replacement of the orthogonality relation is in this case the Parseval relation

$$\int |f(x)|^2\,dx = \frac{1}{2\pi}\int |\tilde{f}(\sigma)|^2\,d\sigma. \tag{1}$$

The geometric meaning of this relation is clear by analogy with Pythagoras' theorem: the square of the length of the vector f is the sum of the squares of the lengths of its Fourier components. Therefore we have indeed some kind of orthogonality for the components.[13]

Completeness for the system of generalized eigenfunctions $\chi_\lambda(x)$ means that each function $\varphi(x)$ can be composed from these generalized eigenfunctions by means of an appropriate integration over the parameter λ, in the same manner as in the theory of Fourier integrals any function can be composed from the exponentials $e^{i\lambda x}$ by means of integration over λ. In our case this fact is a consequence of the general theorem on the reconstruction of an absolutely continuous set function in terms of its derivative. We recall that a completely additive (σ-additive) set function $f(P)$ is said to be *absolutely continuous with respect to the measure* $\sigma(p)$ if $f(p) = 0$ on each set of σ-measure zero, i.e., if $\sigma(P) \to 0$ always implies $f(P) \to 0$. Each function $f(P)$ which is absolutely continuous with respect to the measure $\sigma(P)$ can be represented as an "indefinite integral"

$$f(P) = \int_P \chi\,d\sigma; \tag{2}$$

the function χ is called the *derivative* of the function f with respect to the measure σ.[14]

This allows us to formulate and prove the second fundamental theorem.

Theorem 2. *The set of functionals χ_λ constructed in Theorem 1 is orthogonal and complete in the sense that for each test function $\varphi \in H(e)$ one has the following relations*

$$\varphi = \int_{-\infty}^{\infty} \overline{(\chi_\lambda, \varphi)}\,\chi_\lambda\,d\sigma(\lambda), \tag{3}$$

[13] *Translator's note:* One could also say that although orthogonality has a meaning in this case, normalization leads to divergent integrals. It may be more correct to consider the Parseval identity (1) as a completeness relation, as is usually done by mathematical physicists.

[14] Cf. e.g., S. Saks, "Theory of the Integral," Warszawa-Lwow, 1937.

$$\| \varphi \|^2 = \int_{-\infty}^{\infty} |(\chi_\lambda , \varphi)|^2 \, d\sigma(\lambda). \tag{4}$$

In particular, if $(\chi_\lambda , \varphi) = 0$ for all λ, *then* $\varphi = 0$.

Proof. We make use of the so-called canonical representation by means of a function space of the subspace $H(e)$ generated by the vectors $e_\lambda = E_\lambda e$. We associate to the vector e the function $e(\lambda) \equiv 1$, and to the vector $E(\Delta)e$ we associate the characteristic function of the interval Δ. We then extend this association by making use of linear combinations (of characteristic functions and projection operators) and limits, to all functions of L_σ^2 and the whole space $H(e)$, respectively (L_σ^2 is the Hilbert space of all functions which are square integrable with respect to the measure σ). The association is realized by the formula

$$f = \int_{-\infty}^{\infty} f(\lambda) \, dE_\lambda e \qquad (f(\lambda) \in L_\sigma^2, f \in H(e)), \tag{5}$$

and one can show that for arbitrary $g \in H(e)$

$$(f, g) = \int_{-\infty}^{\infty} f(\lambda) \overline{g(\lambda)} \, d\sigma(\lambda) = \int_{-\infty}^{\infty} f(\lambda) \, d(E_\lambda e, g). \tag{6}$$

These equations show that the association of the spaces $H(e)$ and L_σ^2 is in fact an isomorphism (of Hilbert spaces).

We now prove the completeness of the system of eigenfunctionals χ_λ, i.e., the fact that $(\chi_\lambda , \varphi) = 0$ for all λ implies $\varphi = 0$. From the inequality

$$\left|\left(\int_P dE_\lambda e, \varphi\right)\right|^2 \leqslant \left(\int_P dE_\lambda e, \int_P dE_\lambda e\right) (\varphi, \varphi),$$

with P an arbitrary σ-measurable set, it follows that the set function $(\int_P dE_\lambda e, \varphi) = \int_P d(E_\lambda e, \varphi)$ is absolutely continuous with respect to the measure $\sigma(P) = \int_P d(E_\lambda e, e)$: i.e., if $\sigma(P) = \int_P d(E_\lambda e, e) = 0$ then $\int_P d(E_\lambda e, \varphi) = 0$. Indeed

$$\begin{aligned}\left(\int_P dE_\lambda e, \int_P dE_\lambda e\right) &= \int_P d\left(E_\lambda e, \int_P dE_\lambda e\right) \\ &= \int_P d\left(\int_P d(E_\lambda e, E_\lambda e)\right) \\ &= \int_P d\left(\int_P d(E_\lambda e, e)\right) = 0,\end{aligned}$$

whence $(\int_P dE_\lambda e, \varphi) = 0$. Therefore, according to Eq. (1) of Section 4.3 and Eq.(3) above, we have, for arbitrary $f, \varphi \in H(e)$

$$(f, \varphi) = \int_{-\infty}^{\infty} f(\lambda)\, d(E_\lambda e, \varphi) = \int_{-\infty}^{\infty} f(\lambda)(\chi_\lambda, \varphi)\, d\sigma(\lambda).$$

But, on the other hand, according to Eq. (6) with $\varphi \in H(e)$

$$(f, \varphi) = \int_{-\infty}^{\infty} f(\lambda)\, \overline{\varphi(\lambda)}\, d\sigma(\lambda);$$

since $f(\lambda) \in L_\sigma^2$ is arbitrary, then, almost everywhere with respect to σ

$$(\chi_\lambda, \varphi) = \overline{\varphi(\lambda)}, \qquad \text{i.e.} \qquad (\varphi, \chi_\lambda) = \varphi(\lambda). \tag{7}$$

One can say that $\varphi(\lambda)$ is a "Fourier coefficient" in the decomposition of the test function φ with respect to the generalized functions χ_λ.

Replacing f by φ one can rewrite Eq. (5) in the form

$$\varphi = \int_{-\infty}^{\infty} \overline{(\chi_\lambda, \varphi)}\, dE_\lambda e = \int_{-\infty}^{\infty} \overline{(\chi_\lambda, \varphi)}\, \chi_\lambda\, d\sigma(\lambda),$$

which coincides with the required formula (3). Equations (6) and (7) lead to the Parseval identity

$$(\varphi, \psi) = \int_{-\infty}^{\infty} (\chi_\lambda, \varphi)\, \overline{(\chi_\lambda, \psi)}\, d\sigma(\lambda);$$

which, in particular, for $\varphi = \psi$ becomes

$$\| \varphi \|^2 = \int_{-\infty}^{\infty} |(\chi_\lambda, \varphi)|^2\, d\sigma(\lambda),$$

i.e., the required Eq. (4). This completes the proof of the theorem.

Remark 1. The space H can always be represented as a direct sum of spaces $H(e_\alpha)$, where α belongs to some index set. In each space $H(e_\alpha)$, Theorem 2 establishes the existence of a complete set of eigenfunctionals $\chi_\lambda^{(\alpha)}$. Summing the corresponding identities (4) over all indices α we obtain for an arbitrary $\varphi \in \Phi$ a Parseval identity

$$\| \varphi \|^2 = \sum_\alpha \int_{-\infty}^{\infty} |(\chi_\lambda^{(\alpha)}, \varphi)|^2\, d\sigma_\alpha(\lambda). \tag{8}$$

Correspondingly, one obtains the representation for the vector[15] φ

$$\varphi = \sum_{\alpha} \int_{-\infty}^{\infty} \overline{(\chi_{\lambda}^{(\alpha)}, \varphi)}\, \chi_{\lambda}^{(\alpha)}\, d\sigma_{\alpha}(\lambda). \tag{9}$$

Remark 2. Since in the canonical representation the characteristic function of the interval Δ is associated to the vector $E(\Delta)e$, one can reconstruct the vectors $E(\Delta)e$ in terms of the functionals χ_{λ} by means of integration:

$$E(\Delta)e = \int_{\Delta} \chi_{\lambda}\, d\sigma(\lambda).$$

We see thus that although the generalized eigenfunctions χ_{λ} are elements of the dual space Φ' their appropriately weighted integrals[16] with respect to σ_{λ} (i.e., integrals over sets of positive measure σ_{λ}) realize "concrete" elements of the Hilbert space H, to wit, elements of the form $E(\Delta)e$.

5. Generalized Eigenfunctions of Self-Adjoint Operators

5.1. The Fundamental Theorem

We now apply the general results of Section 4 in order to prove the existence and completeness of the system of generalized eigenfunctions for self-adjoint operators.

As before, we assume that a continuous (in the original topology) inner product (φ, ψ) is defined on the (perfect) nuclear space Φ. As before, we denote the Hilbert space obtained by completion of Φ in the inner product (φ, ψ) by H. Each element $g \in H$ generates a linear functional $f_g[\varphi] = (g, \varphi)$, defined for all $\varphi \in \Phi$. In the sense explained before we have the sequence of inclusion mappings

$$\Phi \subset H \subset \Phi',$$

where Φ' is the dual space of Φ.

[15] *Translator's note:* Cf. in this connection the discussion of direct integrals of Hilbert spaces in Section 4.4, Chapter I, Volume 4.

[16] *Translator's note:* For the reader familiar with quantum mechanics, we mention the relation with Weyl's eigendifferentials, and the physicists' use of wave packets in scattering theory. Cf. also the discussion of Nikodym's method for the treatment of continuous spectra in G. Ludwig, "Grundlagen der Quantenmechanik," pp. 75 ff. Springer, Berlin, 1954.

Let A be a linear operator defined on Φ, and mapping Φ into itself, which is symmetric with respect to the inner product (φ, ψ):

$$(A\varphi, \psi) = (\varphi, A\psi). \tag{1}$$

Such an operator is automatically continuous in Φ. Indeed, let $\varphi_\nu \to \varphi$ then for any $\psi \in \Phi$ we have $(A\varphi_\nu, \psi) = (\varphi_\nu, A\psi) \to (\varphi, A\psi) = (A\varphi, \psi)$, i.e., the sequence $A\varphi_\nu$ converges weakly; but since the nuclear space Φ is perfect this sequence converges also strongly to the element $A\varphi$, i.e., A is continuous.

The adjoint operator A^* is continuous and bounded in the dual space Φ'. Since $\Phi' \supset \Phi$ the operator A^* is also defined on the space Φ, where it coincides with the operator A. Indeed, for $\varphi \in \Phi$ and any $\psi \in \Phi$ we have:

$$(A^*\varphi, \psi) = (\varphi, A\psi) = (A\varphi, \psi),$$

so that $A^*\varphi$ coincides with the element $A\varphi$.

Thus the operator A^* is an extension of the symmetric operator A to the space Φ', so that we can omit in the sequel the sign * from A^* and the operator A is a symmetric operator in Φ.

A linear functional $\chi_\lambda \in \Phi'$ will be called an *eigenfunctional* or a *generalized eigenfunction* (some authors use "eigendistribution") of the operator A, belonging to the eigenvalue λ if

$$A\chi_\lambda = \lambda\chi_\lambda .$$

Theorem 1. *A symmetric linear operator A, defined on the space Φ, which admits a self-adjoint extension to the Hilbert space H, possesses a complete system of eigenfunctionals χ_λ belonging to the dual space Φ'.*

Proof. By assumption and on the basis of the fundamental spectral theorem for self-adjoint operators,[17] the operator A possesses a spectral family $E_\lambda = E(\Delta^\lambda_{-\infty})$. We choose an arbitrary vector $e \in H$ and consider the subspace $H(e)$ spanned by the vectors $e_\lambda = E_\lambda e$. According to Theorem 2 in Section 4, there exists a complete system of functionals

$$\chi_\lambda = \frac{dE_\lambda e}{d(E_\lambda e, e)} .$$

We check that the functionals χ_λ are eigenfunctionals of the operator A. We denote by Δ the interval $[\alpha, \beta]$ containing the point λ, and by

[17] See, e.g., F. Riesz and B. Szökefalvy-Nagy, Chapter 8, Section 2. (Ref. 3, p. 169, this book.)

$E(\Delta)$ the operator $E_\beta - E_\alpha$. Let the interval Δ contract to the point λ. For any $\varphi \in \Phi$ we have, almost everywhere with respect to the measure σ

$$\begin{aligned}(A\chi_\lambda, \varphi) = (\chi_\lambda, A\varphi) &= \lim \left(\frac{E(\Delta)e}{\sigma(\Delta)}, A\varphi\right) \\ &= \lim \frac{1}{\sigma(\Delta)} \left(E(\Delta)e, \int_{-\infty}^{\infty} \lambda \, dE_\lambda \, \varphi\right) \\ &= \lim \frac{1}{(\sigma\Delta)} \left(e, \int_{-\infty}^{\infty} \lambda \, dE(\Delta) \, E_\lambda \varphi\right) \\ &= \lim \frac{1}{\sigma(\Delta)} \left(e, \int_{\Delta} \lambda \, dE_\lambda \, \varphi\right) \\ &= \lim \frac{1}{\sigma(\Delta)} \left(\int_{\Delta} \lambda \, dE_\lambda \, e, \varphi\right) \\ &= \lim \left(\int_{\Delta} \frac{\lambda \, dE_\lambda \, e}{\sigma(\Delta)}, \varphi\right) = (\lambda\chi_\lambda, \varphi),\end{aligned}$$

whence

$$A\chi_\lambda = \lambda\chi_\lambda,$$

as required.

As before, the space H can be decomposed into an orthogonal sum of spaces $H(e_\alpha)$. Using the theorem for every $H(e_\alpha)$ and combining the results we obtain the existence (in Φ') of a system of eigenfunctionals $\chi_\lambda^{(\alpha)}$ for the operator A, which is complete in the whole space Φ'.

5.2. Differential Operators Defined in the Whole Space

Example 1. Let A denote a linear differential operator

$$A\varphi = \sum a_n(x) D^k \varphi(x) \tag{1}$$

with real, infinitely differentiable coefficients, defined throughout R_n, and symmetric in the inner product

$$(\varphi, \psi) = \int \overline{\varphi(x)} \, \psi(x) \, dx; \tag{2}$$

This means that for two functions φ and ψ of compact support

$$(A\varphi, \psi) = (\varphi, A\psi). \tag{3}$$

We choose the (perfect) N-space Φ to consist of infinitely differentiable functions (e.g., a space of type $K\{M_p\}$) and such that the operator A maps the space Φ into itself and remains symmetric (i.e., verifies (3)). In addition we assume that the space Φ is densely included in the space $L_2(R_n)$ of all square-integrable functions $\varphi(x)$.

Then the operator A can be extended to a self-adjoint operator in $L_2(R_n)$.[18] Consequently the assumptions of Theorem 1 are satisfied and we obtain

Theorem 2. *A differential operator A which satisfies the enumerated conditions possesses a complete system of generalized eigenfunctions $\chi_\lambda(x)$ belonging to the space Φ'.*

If we know the general form of continuous linear functionals over the space Φ we can describe the general form of generalized eigenfunctions of the operator A.

Let us assume, for example, that the space Φ is the space S of functions $\varphi(x)$ which are infinitely differentiable and which together with their derivatives of all orders fall off at infinity faster than any power of $1/|x|$. The space $\Phi = S$ is useful in particular for the treatment of differential operators A for which the coefficients together with their derivatives do not increase faster than a power of $|x|$ at infinity. Any linear continuous functional (distribution) on the space S is a derivative of finite order p of a continuous function which does not increase faster than $|x|^p$. We thus arrive at the conclusion: *an operator A defined on S, which satisfies the enumerated conditions, admits a complete system of generalized eigenfunctions, each of which is a derivative of order p of a continuous function which does not increase at infinity faster than $|x|^p$ (for fixed p).*

5.3. Differential Operators in Regions with Boundaries

Example 2. The first example dealt with an operator A acting on functions defined in the whole space R_n. Here we consider an operator A defined for functions with support in a region with boundary.

Let A be a linear differential operator

$$A\varphi \equiv \sum a_n(x) D^k \varphi(x) \tag{1}$$

[18] F. Riesz and B. Szökefalvy-Nagy, p. 120. (Ref. 3, this book, p. 169.)

with real infinitely differentiable coefficients defined in the region G *of* R_n and symmetric with respect to the inner product

$$(\varphi, \psi) = \int_G \overline{\varphi(x)}\, \psi(x)\, dx. \tag{2}$$

More precisely, we assume that for any two functions, φ and ψ, of compact support, defined (and infinitely differentiable) in the region G, and vanishing in a neighborhood of the boundary Γ of G, we have

$$(A\varphi, \psi) = (\varphi, A\psi).$$

The operator A can be extended to a self-adjoint operator on the space $L_2(G)$.

Its domain Ω_A is a subset of functions $\varphi(x) \in L_2(G)$ satisfying definite *boundary conditions*. The condition that A be self-adjoint means that for any function $\psi(x) \in L_2(G)$ for which the expression $(\psi(x), A\varphi(x))$ is a bounded functional of $\varphi(x)$ on $L_2(G)$, belongs to the domain of the operator A and satisfies the boundary conditions which characterize the domain of this operator.

We assume that A is given as a self-adjoint operator from the outset. Then according to Theorem 1, A will have a complete system of generalized eigenfunctions defined as functionals over a nuclear space Φ which contains all infinitely differentiable functions of compact support and is invariant under the action of A.

The question arises as to the sense in which these generalized eigenfunctions satisfy the boundary conditions, characteristic for the domain of the operator A.

Since the generalized eigenfunctions $\chi_\lambda(x)$ are not in general elements of $L_2(G)$, we have no possibility of considering the expression $(\chi_\lambda(x), A\varphi(x))$ and showing that it is a bounded linear functional of $\varphi(x)$. We proceed in a different manner. If the boundary of the region G is entirely at finite distances, then we have:

Theorem 3. *Let $\chi_\lambda(x)$ be a generalized eigenfunction of an operator A of order p, which in reality is an ordinary function and admits ordinary derivatives up to order p; then the product $g_\lambda(x) = \chi_\lambda(x)\, e(x)$ belongs to the domain of A, where $e(x)$ is an infinitely differentiable function of compact support which is equal to one in a neighborhood of the boundary Γ of G.*

Proof. We show that the functional of

$$(g_\lambda, A\varphi)$$

is bounded in $L_2(G)$.

Interpreting g_λ as a generalized function and φ as a test function, the symmetry of the operator A implies

$$(g_\lambda, A\varphi) = (Ag_\lambda, \varphi).$$

But considered as an operation on a generalized function, the operator A does not always act as the operation $\sum a_k(x)D^k$, even if $g_\lambda(x)$ possesses all the required derivatives. We discuss this problem in general, i.e., we investigate the meaning of applying the operator A to a generalized function defined by an ordinary, p times continuously differentiable, function $f(x)$. If $\varphi(x)$ is a test function of compact support, Green's formula implies

$$\begin{aligned}(f, A\varphi) &= \int_G f \cdot A\varphi \cdot dx \\ &= \int_G A_1 f \cdot \varphi \cdot dx + \int_\Gamma L[f, \varphi]\, d\sigma \\ &= (Af, \varphi),\end{aligned} \tag{3}$$

where $A_1 f$ means the application in the usual sense of the differential operator A to the function $f(x)$, and L is a bilinear form appearing due to the integration by parts. We see that the expressions Af and $A_1 f$ are in general different, the difference depending on the boundary values of the function f.

We show, however, that for a generalized eigenfunction these expressions coincide.

If $f = \chi_\lambda$ is an eigenfunctional (generalized eigenfunction) of the operator A, we have

$$(f, A\varphi) = (Af, \varphi) = \lambda(\chi_\lambda, \varphi) = \lambda \int_G \chi_\lambda(x)\varphi(x)\, dx. \tag{4}$$

Comparing (3) and (4) we see that

$$\int_\Gamma L[f, \varphi]\, d\sigma = \int_G [\lambda\chi_\lambda(x) - A_1 f]\varphi\, dx. \tag{5}$$

But the functional on the left-hand side is singular whereas the functional in the right-hand side is regular, which implies that both have to vanish. Indeed, if φ is chosen so as to vanish along the boundary, we obtain

$$\int_G [\lambda\chi_\lambda(x) - A_1\chi_\lambda]\varphi\, dx = 0$$

and since φ is otherwise arbitrary, we have everywhere except the boundary Γ

$$\lambda\chi_\lambda(x) - A_1\chi_\lambda = 0.$$

It follows that for an *arbitrary* φ the right-hand side of (5) vanishes and hence the left-hand side vanishes too, for arbitrary φ. Consequently $\chi_\lambda(x)$ satisfies the equation

$$A_1\chi_\lambda(x) = \lambda\chi_\lambda(x),$$

i.e., it is an ordinary eigenfunction of A. Further, for any test function

$$\int_\Gamma L[\chi_\lambda, \varphi]\, d\sigma = 0,$$

whence

$$\int_\Gamma L[\chi_\lambda e, \varphi]\, d\sigma = 0,$$

since χ_λ and $\chi_\lambda e$ coincide in the neighborhood of the boundary. Consequently

$$(e(x)\chi_\lambda(x), A\varphi) = \int_G A_1 e(x)\chi_\lambda(x) \cdot \varphi(x)\, dx;$$

therefore, this functional of φ is bounded, and remains bounded throughout the space H, since the function $A_1 e(x)\chi_\lambda(x)$ is of compact support if $e(x)\chi_\lambda(x)$ is. Therefore $e(x)\chi_\lambda(x)$ is in the domain of the operator A and in particular, satisfies the boundary conditions. This completes the proof.

We now give the following general definition.

Definition. A functional f is said to be a solution of the equation $Bf = 0$, satisfying the boundary conditions, if for all infinitely differentiable functions $\varphi(x)$ in the domain of the self-adjoint operator B, it satisfies the equation

$$(f, B\varphi) = 0.$$

We now prove that the generalized eigenfunctions χ_λ of the self-adjoint operator A satisfy the boundary conditions in the sense of this definition. Here $B = A - \lambda E$. Let $\varphi(x)$ belong to the domain of the operator A. Then

$$(AE(\Delta)e, \varphi) = \int_\Delta \lambda\, d(E_\lambda e, \varphi),$$

and the functional $E(\Delta)e$ satisfies the boundary conditions. Therefore

$$(E(\Delta)e, A\varphi) = \int_\Delta \lambda\, d(E_\lambda e, \varphi) = \int_\Delta \lambda(\chi_\lambda, \varphi)\, d\sigma(\lambda),$$

consequently

$$\left(\frac{dE_\lambda e}{d\sigma}, A\varphi\right) = (\chi_\lambda, A\varphi) = \lambda(\chi_\lambda, \varphi),$$

whence

$$(\chi_\lambda, (A - \lambda E)\varphi) = 0,$$

as asserted.

5.4. The Sturm-Liouville Operator

Example 3. We consider the operator

$$Ay = -y'' + q(x)y \qquad (0 \leqslant x < \infty, \quad \operatorname{Im} q(x) = 0) \tag{1}$$

with infinitely differentiable coefficient $q(x)$. Under certain conditions to be imposed on its domain (and on the coefficient $q(x)$), the operator is self-adjoint.[19] In particular, this is true for the boundary condition

$$y'(0) = \theta y(0) \qquad (\theta \text{ real}) \tag{2}$$

Theorem 1 implies the existence of a *complete system of generalized eigenfunctions* $\chi_\lambda(x)$ each of which satisfies the equation

$$-y'' + q(x)y = \lambda y. \tag{3}$$

We note that in the case under consideration these eigenfunctions are ordinary infinitely differentiable functions, since an ordinary differential equation without singular points does not admit other kind of solutions in the class of generalized functions (cf. Chapter I, Volume 1).

We saw in Example 2, that such eigenfunctions satisfy the boundary conditions in the usual sense.

We now formulate for the special case at hand the theorem on completeness of the system of eigenfunctions (Theorem 2 in Section 4):

[19] Cf. M. A. Naimark, "Lineinie differentsialnye operatory" (Linear Differential Operators). Gostekhizdat, Moscow, 1954 (German Ed., Berlin, 1960).

Theorem 4. *Let e_α denote the set of generating vectors spanning the space H for the operator $Ay = -y'' + q(x)y$, and let $\sigma_\alpha(\lambda)$ denote the corresponding monotone functions: $\sigma_\alpha(\lambda) = (E_\lambda e_\alpha, e_\alpha)$. The eigenfunction $\chi_\alpha(\lambda)$ obtained by differentiating the vector $E_\lambda e_\alpha$ with respect to the measure $\sigma_\alpha(\lambda)$ is a solution of Eq. (3) and can therefore differ at most by a numerical factor $b_\alpha(\lambda)$ from the (unique) classical solution $y(x, \lambda)$ defined by the initial conditions $y(0, \lambda) = 1$, $y'(0, \lambda) = \theta$:*

$$\chi_\alpha(\lambda) = b_\alpha(\lambda)y(x, \lambda).$$

The expansion formula for a function $\varphi(x)$ of compact support takes the form

$$\varphi(x) = \sum_\alpha \int_0^\infty \overline{\left[\int_0^a b_\alpha(\lambda)y(\xi, \lambda)\varphi(\xi)\, d\xi\right]}\, y(x, \lambda)b_\alpha(\lambda)\, d\sigma_\alpha(\lambda)$$

$$= \int_0^\infty y(x, \lambda)\left[\int_0^a y(\xi, \lambda)\,\overline{\varphi(\xi)}\, d\xi\right]\sum_\alpha b_\alpha^2(\lambda)\, d\sigma_\alpha(\lambda). \tag{4}$$

We denote the sum $\sum b_\alpha^2(\lambda)\, d\sigma_\alpha(\lambda)$ by $d\sigma(\lambda)$; $\sigma(\lambda)$ is the spectral function of the problem. The function

$$F(\lambda) = \int_0^a y(\xi, \lambda)\,\overline{\varphi(\xi)}\, d\xi = (\chi_\lambda, \varphi) \tag{5}$$

will be called the *Fourier-Sturm-Liouville transform* of the function $\varphi(x)$. Eq. (4) shows that the function $\varphi(x)$ can be obtained from its Fourier-Sturm-Liouville transform by means of the inversion formula

$$\varphi(x) = \int_{-\infty}^\infty y(x, \lambda)F(\lambda)\, d\sigma(\lambda). \tag{6}$$

The *Parseval identity becomes*

$$\|\varphi\|^2 = \int_0^a |\varphi(x)|^2\, dx = \int_{-\infty}^\infty |F(\lambda)|^2\, d\sigma(\lambda). \tag{7}$$

Our derivation of this formula is valid for functions $\varphi(x)$ of compact support.

Here the function $F(\lambda)$ is an element of the Hilbert space L_σ^2 of σ-square-integrable functions. Equations (5) and (6) show that the correspondence between the functions $\varphi(x)$ and $F(\lambda)$ can be extended to a correspondence between all functions $\varphi(x)$ belonging to the Hilbert space $L_2(0, \infty)$ on the one hand, and the functions $F(\lambda)$ of the space L_σ^2

on the other hand. The correspondence is a linear norm-preserving map (unitary transformation) and is therefore a Hilbert-space isomorphism.

Note. A similar theorem holds, of course, for the nth order self-adjoint operator

$$Ay \equiv y^{(n)}(x) + a_1(x)y^{(n-1)}(x) + \cdots + a_n(x)y(x)$$

with infinitely differentiable coefficients $a_j(x)$.

5.5. The System of Simultaneous Generalized Eigenfunctions of a Pair of Self-Adjoint Operators

The following theorem is analogous to the theorem of simultaneous reduction to a canonical form of two quadratic forms.

Example 4. Let A and B denote self-ajdoint operators defined in the (perfect) nuclear space Φ; let further B be positive definite in the inner product (φ, ψ), i.e., $(B\varphi, \psi) > 0$ for $\varphi \neq 0$ and invertible in the Hilbert space H (i.e., B^{-1} is defined there).

We show that *there exists a complete system of generalized functions* χ_λ *satisfying the conditions*

$$A\chi_\lambda = \lambda B\chi_\lambda . \tag{1}$$

The operator $B^{-1}A$ is selfadjoint with respect to the inner product $(B\varphi, \psi)$, since

$$(B(B^{-1}A\varphi), \psi) = (A\varphi, \psi) = (\varphi, A\psi) = (B\varphi, B^{-1}A\psi).$$

The classical spectral theorem implies the existence of a spectral family E_λ for the operator $B^{-1}A$. Differentiating this function with respect to the measure σ_λ, where $\sigma_\lambda(p) = \int_p d(E_\lambda e, e)$, we obtain on the basis of the theorems of Section 4 the complete family of functionals

$$\chi_\lambda = \frac{dE_\lambda e}{d\sigma_\lambda} \in \Phi'.$$

It remains to be shown that these functionals are solutions of Eq. (1).

From the equation

$$B^{-1}AE(\Delta) = \int_\lambda^{\lambda'} \mu \, dE_\mu \qquad (\Delta = (\lambda, \lambda'))$$

we derive

$$AE(\Delta) = B\int_{\lambda}^{\lambda'} \mu \, dE_\mu = \int_{\lambda}^{\lambda'} \mu \, d(BE_\mu),$$

hence

$$\frac{AE(\Delta)e}{\sigma(\Delta)} = \frac{1}{\sigma(\Delta)}\int_{\lambda}^{\lambda'} \mu \, d(BE_\mu).$$

Taking the limit $\lambda' \to \lambda$ we obtain Eq. (1).

As an example we consider the equation proposed by S. L. Sobolev and R. A. Alexandryan:

$$\frac{\partial^2}{\partial t^2}\left(\frac{\partial^2 u}{\partial x^2} + \frac{\partial^2 u}{\partial y^2}\right) = \frac{\partial^2 u}{\partial x^2}. \tag{2}$$

with the following boundary and initial conditions: the function $u(x, y, t)$ takes given values on the smooth boundary Γ of the region G and $u(x, y, 0) = f(x, y)$, with f a given function.

Separation of variables leads to the following equation for the t-independent function v

$$\frac{\partial^2 v}{\partial x^2} - \lambda\left(\frac{\partial^2 v}{\partial x^2} + \frac{\partial^2 v}{\partial y^2}\right) = 0, \tag{3}$$

where v is subjected to boundary conditions on Γ. The Laplace operator $\Delta = (\partial^2/\partial x^2) + (\partial^2/\partial y^2)$ is known to be positive definite. In addition, both operators $\partial^2/\partial x^2$ and Δ, when defined on functions $\varphi(x)$ which vanish on the boundary Γ are obviously self-adjoint. The theorem stated above leads then to the result:

Theorem 5. *The solutions of Eq.* (2) *with given values on the boundary form a complete system of generalized eigenfunctions.*

5.6. The Case of Coefficients of Finite Order of Differentiability

The condition that the coefficients of the differential operators A considered above be infinitely differentiable (which was imposed due to the infinite differentiability of the test functions of the space Φ, so that A maps Φ into itself) is often too restrictive for problems in the theory of differential operators.

The following simple consideration should allow us to weaken this condition. It is clear that the spectral family E_λ and the complete system of functionals χ exists also for an operator A which does not necessarily map Φ into itself. It is sufficient that the operator A be defined on a dense subspace $L_A \subset \Phi$ and that it maps this subspace into a Hilbert space H, where it admits a self-adjoint extension.

But the equation $A\chi_\lambda = \lambda\chi_\lambda$ may in general lose its meaning in this case, since A is no longer defined on the dual space Φ'.

There is an important case in which this theorem remains valid. Let us assume that the operator A maps the space Φ not simply into H, but into one of the normed spaces $\Phi_p \subset H$. Then the adjoint operator A^* maps the space Φ_p' into Φ'. But, as we have seen before, the operator A^* is an extension of A, so that the * can be omitted. We see that our assumption is equivalent to the fact that A is defined on the normed space Φ_p. Let further the functionals $\chi_\lambda = dE_\lambda e/d\sigma_\lambda$ also belong to the space Φ_p'. Then the expression $A\chi_\lambda$ is meaningful and, as has been shown $A\chi_\lambda = \lambda\chi_\lambda$ as required. Similar considerations can be used for linear differential operators.

We assume that the index p characterizing the order of the functional over Φ_p coincides with the maximal number of continuous derivatives which the functions making up that space admit. (This is certainly true for the previously considered test function spaces K, S, $K\{M_p\}$.)

If this assumption is satisfied, the multiplication by a function $a(x)$ which admits continuous derivatives up to order p (assuming that the behavior at infinity does not disagree with this multiplication) is an operation mapping the space Φ into Φ_p. Therefore the linear differential operator

$$A = \sum a^k D^k$$

with coefficients $a_k(x)$ of class C^p, will map the space Φ into Φ_p.

This remark allows one to replace in the preceding examples the infinite differentiability of the coefficients by a condition of sufficient smoothness, i.e., existence of derivatives up to a certain order only. The question remains however, how high to take this order, since it also depends on the index p for which the vector $E_\lambda e$ is of bounded variation when considered a functional over the space with the norm $\| \; \|_p$.

We shall see that in the case of the ordinary inner product $(\varphi, \psi) = \int \varphi\bar{\psi}\, dx$ it suffices to require the existence of first derivatives only. Further, it is often possible to show that the generalized eigenfunctions are in fact ordinary functions. It will also be possible to indicate the growth caracteristics of these functions at infinity.

6. The Structure of the Generalized Eigenfunctions

6.1. The Fundamental Theorem

In this section we investigate the structure of the generalized eigenfunctions $\chi_\lambda = dE_\lambda e/d\sigma_\lambda$ under the assumption that the operators E_λ act in the ordinary Hilbert space of square-integrable functions $\varphi(x)$ $(x \in R_n)$ with the inner product

$$(\varphi, \psi) = \int \varphi(x)\overline{\psi(x)}\, dx. \tag{1}$$

The test function space will be assumed, as before, to be perfect (we need not assume here the nuclearity of Φ).

Furthermore it is assumed that the first two norms $\| \varphi \|_0$ and $\| \varphi \|_1$ of the space Φ are of a special form, to wit:

(1) $\| \varphi \|_0^2 = \int | \varphi(x)|^2 M_0(x)\, dx$ with $1/M_0(x)$ a square-integrable funtion, and

(2) $\| \varphi \|_1 = \max\, [| D\varphi(x)| \cdot M_1(x)]$, with $(| x_1 \cdots x_n |)^{1/2}/M_1(x)$ square-integrable and $D = \partial^n/\partial x_1 \cdots \partial x_n$.

The first assumption implies that all test functions $\varphi(x)$ are square-integrable, whereas the second assumption implies that the functions $D\varphi(x)$ multiplied by $(| x_1 \cdots x_n |)^{1/2}$ remain square-integrable. Since the remaining norms are arbitrary, many problems with operators which are symmetric with respect to the ordinary scalar product can be included in the treatment presented here.

We denote the completions of Φ with respect to the norms $\| \ \|_0$ and $\| \ \|_1$ by Φ_0 and Φ_1 respectively.

Theorem 1. *Let the self-adjoint operator A be defined on the Hilbert space H with inner product* (1). *Then the functionals $\chi_\lambda = dE_\lambda e/d\sigma_\lambda$ are defined and continuous on the space Φ_1.*

Proof. According to Section 2.2, it suffices to show that the expressions

$$\sum \| E_{\lambda_{j+1}} e - E_{\lambda_j} e \|_{\Phi_1'},$$

are uniformly bounded. We define $E(\Delta_j) = E_{\lambda_{j+1}} - E_{\lambda_j}$. For different j the operators $E(\Delta_j)$ are mutually orthogonal, and for any vector e, $\| e \| = 1$:

$$\sum_j (E(\Delta_j)e, E(\Delta_j)e) = (e, e) = 1.$$

The action of the functionals $E(\Delta_j)e$ on the test function is described by

$$(E(\Delta_j)e, \varphi) = \int E(\Delta_j)e(x)\, \overline{\varphi(x)}\, dx. \tag{2}$$

We define the function $G_{\Delta_j}(x)$:

$$G_{\Delta_j}(x) \equiv G_{\Delta_j}(x_1, \ldots, x_n) = \int_0^{x_1} \cdots \int_0^{x_n} E(\Delta_j)e(\xi)\, d\xi_1 \cdots d\xi_n\,. \tag{3}$$

The function $E(\Delta_j)e(x)$ can be obtained from $G_{\Delta_j}(x)$ by means of differentiation:

$$E(\Delta_j)e(x) = DG_{\Delta_j}(x),$$
$$D = \frac{\partial^n}{\partial x_1 \cdots \partial x_n}\,. \tag{4}$$

Using the Schwarz inequality, one can estimate the growth of $G_{\Delta_j}(x)$:

$$|\, G_{\Delta_j}(x_1, \ldots, x_n)| \leqslant (|\, x_1 \cdots x_n\, |)^{1/2} \left(\int_0^{x_1} \cdots \int_0^{x_n} |\, E(\Delta_j)e(\xi)|^2\, d\xi\right)^{1/2}$$
$$\leqslant (|\, x_1 \cdots x_n\, |)^{1/2}, \tag{5}$$

under the assumption that $e(x)$ is normalized.

We show that

$$\int E(\Delta_j)e(x)\, \overline{\varphi(x)}\, dx = (-1)^n \int G_{\Delta_j}e(x)\, \overline{D\varphi(x)}\, dx. \tag{6}$$

For a function of compact support $\varphi(x)$ the validity of (6) follows directly by integration by parts (the integrated terms vanish due to the support property of φ). For an arbitrary test function $\varphi(x)$ one can always form a sequence $\{\varphi_\nu(x)\}$ of functions of compact support in Φ which converge to $\varphi(x)$ in the topology of that space (cf. Section 4, Chapter II, Volume 2). Then the sequence of derivatives $\{D\varphi_\nu\}$ converges to $D\varphi$ in Φ. As before we assume that convergence in the topology of Φ implies weak convergence in H (i.e., with respect to the inner product). Thus one can take the limit $\nu \to \infty$ under the integral in

$$\int E(\Delta_j)e(x)\, \overline{\varphi_\nu(x)}\, dx = (-1)^n \int G_{\Delta_j}e(x)\, \overline{D\varphi_\nu(x)}\, dx$$

which yields the Eq. (6).

Consequently

$$(E(\Delta_j)e, \varphi) = (-1)^n \int G_{\Delta_j}e(x)\, \overline{D\varphi(x)}\, dx.$$

whence

$$\begin{aligned}|(E(\Delta_j)e, \varphi)| &\leqslant \int |G_{\Delta_j}e(x)D\varphi(x)|\, dx \\ &\leqslant \int \frac{|G_{\Delta_j}e(x)M_1(x)D\varphi(x)|}{M_1(x)}\, dx \\ &\leqslant \sup_x M_1(x)\, |D\varphi(x)| \cdot \int \frac{|G_{\Delta_j}e(x)|}{M_1(x)}\, dx.\end{aligned}$$

Thus

$$\| E_{\Delta_j}e \|_{\Phi_1'} \leqslant \int \frac{|G_{\Delta_j}e(x)|}{M_1(x)}\, dx,$$

and consequently

$$\sum \| E_{\Delta_j}e \|_{\Phi_1'} \leqslant \int \frac{\sum |G_{\Delta_j}e(x)|}{M_1(x)}\, dx.$$

We have ($|\epsilon_j(x)| = 1$):

$$\begin{aligned}\sum |G_{\Delta_j}e(x)| &= \int_0^x \sum_j \epsilon_j(\xi)E(\Delta_j)e(\xi)\, d\xi \\ &\leqslant (|x_1 \cdots x_n|)^{1/2} \left(\int_0^x \left| \sum_j \epsilon_j(\xi)E(\Delta_j)e(\xi) \right|^2 d\xi\right)^{1/2} \\ &\leqslant (|x_1 \cdots x_n|)^{1/2} \left(\int_R \left| \sum_j \epsilon_j(\xi)E(\Delta_j)e(\xi) \right|^2 d\xi\right)^{1/2} \\ &\leqslant (|x_1 \cdots x_n|)^{1/2} \\ &\quad \times \left(\int_R \left[\sum_j \epsilon_j(x)E(\Delta_j)e(\xi)\right]\left[\sum_k \overline{\epsilon_k(x)E(\Delta_k)e(\xi)}\right] d\xi\right)^{1/2} \\ &= (|x_1 \cdots x_n|)^{1/2} \left(\int_R \sum_{j,k} \epsilon_j(\xi)\, \overline{\epsilon_k(\xi)} E(\Delta_j)e(\xi)\, \overline{E(\Delta_k)e(\xi)}\, d\xi\right)^{1/2} \\ &= (|x_1 \cdots x_n|)^{1/2} \left(\sum_{j,k} \epsilon_j(\xi)\, \overline{\epsilon_k(\xi)} \int_R E(\Delta_j)e(\xi)\, \overline{E(\Delta_k)e(\xi)}\, d\xi\right)^{1/2} \\ &= (|x_1 \cdots x_n|)^{1/2} \left(\sum_j (E(\Delta_j)e, E(\Delta_j)e)\right)^{1/2} = (|x_1 \cdots x_n|)^{1/2},\end{aligned}$$

since the vector e is assumed normalized.

As a result we obtain

$$\sum \| E_{\Delta_j} e \|_{\Phi_1'} \leqslant \int \frac{(|x_1 \cdots x_n|)^{1/2}}{M_1(x)} \, dx = C < \infty, \tag{7}$$

as required.

The general form of a linear continuous functional on the space Φ_1 is known (cf. Section 4, Chapter II, Volume 2). It is given by

$$(f, \varphi) = \int M_1(x) f(x) D\varphi(x) \, dx, \tag{8}$$

where $f(x)$ is a bounded measurable function. In other words the functional f is obtained by letting $D = \partial^n/\partial x_1 \cdots \partial x_n$ act on the function $g(x) = M_1(x) f(x)$ which does not increase faster than $CM_1(x), f(x)$ being bounded.

For $M_1(x)$ one can choose any function, as long as condition 2 is satisfied. For example, the function

$$M_1(x) = (1 + |x|)^{(3n/2)+\epsilon}.$$

satisfies all the requirements. We thus reach the conclusion:

Theorem 2. *The generalized eigenfunctions $\chi_\lambda(x)$ of any self-adjoint operator A defined on the space L_2 are derivatives (of the type $\partial^n/\partial x_1 \cdots \partial x_n$) of measurable functions which do not increase faster than $(1 + |x|)^{(3n/2)+\epsilon}$.*

6.2. The Case of a Differential Operator

In Section 5 we have considered the problem of eigenfunction expansions of a differential operator. In Sections 5.2–5.5 it was assumed that the coefficients of the operator are infinitely differentiable functions in order to maintain invariance of the space Φ under the action of the operator A, i.e., in order that $\varphi \in \Phi$ imply $A\varphi \in \Phi$. Theorem 1 of Section 6.1 allows us to lift this restriction.

Let A be a differential operator with coefficients $a_j(x)$ having continuous derivatives $Da_j(x) = \partial^n a_j(x)/\partial x_1 \cdots \partial x_n$ and satisfying the inequality

$$|Da_j(x)| \leqslant M(x), \tag{1}$$

where $M(x)$ is a monotone function.

We consider the function spaces Φ_0 and Φ_1 introduced in the preceding section. Let the functions $M_0(x)$ and $M_1(x)$ which fix the norms in these spaces be such that

$$\frac{M(x)(|x_1 \cdots x_n|)^{1/2}}{M_1(x)} \quad \text{and} \quad \frac{M(x)}{M_0(x)} \quad \text{are integrable.} \tag{2}$$

Taking into account the statement at the end of Section 5, it results that

$$\frac{d(E_\lambda e, A\varphi)}{d\sigma_\lambda} = (\chi_\lambda, A\varphi)$$

exists and that

$$(\chi_\lambda, A\varphi) = \lambda(\chi_\lambda, \varphi).$$

With this we have proved the following

Theorem 3. *Any self-adjoint differential operator with coefficients having a continuous derivative admits a complete system of generalized eigenfunctions which are linear functionals over the space* Φ_1.

Theorem 2 makes it possible to get some information about the behavior of generalized eigenfunctions (therefore also for eigenfunctions) for differential operators A defined in the whole space. On the other hand, if the domain of the operator A consists of functions defined in a bounded region G, then, repeating the arguments of the proof of Theorem 1 one can see that the eigenfunctionals $\chi_\lambda(x)$ are derivatives of *bounded* continuous functions $f_\lambda(x)$. It should be noted that the operator A can also be singular, i.e., a differential operator with coefficients having singular points.

Any generalized solution of a Sturm-Liouville problem $\chi_\lambda = dE_\lambda e/d\sigma_\lambda$ is also an ordinary solution (Section 5), i.e., a function $y(x, \lambda)$ which is twice continuously differentiable and satisfies the equation

$$-y'' + q(x)y = \lambda y.$$

For any function of compact support $\varphi(x)$ we have

$$(\chi_\lambda, \varphi) = \int_a^b f_\lambda(x)\varphi'(x)\,dx = \int_a^b y(x, \lambda)\varphi(x)\,dx, \tag{3}$$

with $f_\lambda(x)$ a measurable function. Equation (3) implies the inequality

$$\left| \int_0^x y(\xi, \lambda)\, d\xi \right| < C; \tag{4}$$

and consequently we have:

Theorem 4. *The eigenfunctions $y(x, \lambda)$ of a Sturm-Liouville equation over a finite interval satisfy an inequality* (4) *independent of whether $q(x)$ is or is not singular. In other words the eigenfunctions are integrable, even in the presence of singularities.*

Note. A similar theorem can be derived also for the case of a self-adjoint eigenvalue equation of order $2n$, defined on a finite interval $[a, b]$:

$$y^{2n} + q_1(x)y^{(2n-1)} + \cdots + q_{2n}(x)y = \lambda y.$$

7. Dynamical Systems[20]

We consider the system of ordinary differential equations

$$\begin{aligned} \frac{dy_1}{dt} &= Y_1(y_1, \ldots, y_n), \\ &\cdots\cdots\cdots \\ &\cdots\cdots\cdots \\ \frac{dy_n}{dt} &= Y_n(y_1, \ldots, y_n), \end{aligned} \tag{1}$$

or, in vector notation,

$$\frac{dy}{dt} = Y(y), \tag{1'}$$

where the point $y = (y_1, \ldots, y_n)$ belongs to a C^∞-manifold $\mathfrak{M}$, and the functions $Y_1, \ldots, Y_n$ are such that the existence and uniqueness of the solution of the system (1) is guaranteed for arbitrary initial data $y^0 = (y_1^0, \ldots, y_n^0)$ on the manifold $\mathfrak{M}$.

Physically, one can interpret the system (1) as the equations of motion of points on the manifold $\mathfrak{M}$. For any y^0 and any value of t one can

[20] Cf. V. V. Nemytskii and V. V. Stepanov, "Kachestvennaya teoriya differentsialnykh uravnenii" (Qualitative Theory of Differential Equations) 2nd Russian Ed. Gostekhizdat M.-L. 1949.

construct the point y_t as a solution of (1) satisfying the initial condition $y(0) = 0$; y_t describes the position of the point y at the "instant" t.

Thus the system (1) defines a transformation of the manifold $\mathfrak{M}$ into itself

$$Q_t y_0 = y_t .$$

These transformations obviously form a group

$$Q_{t+s} = Q_t \cdot Q_s .$$

One can apply the operator Q_t not only to individual points on the manifold, but to a whole region G. By $Q_t G = G_t$ one should understand the set of all points $y(t)$ at time t which at time 0 are situated inside the region G.

The system (1) should admit an "integral invariant", i.e., we assume there exists a function $F(x) > 0$, such that the "volume" defined by the differential form $F(y)\, dy_1 \cdots dy_n$

$$V(G) = \int_G F(y_1, \ldots, y_n)\, dy_1 \cdots dy_n \geqslant 0, \tag{2}$$

is invariant under the transformation Q_t (or as one usually says: invariant under the motion of the system):

$$V[G_t] = V[G]. \tag{3}$$

In the dynamical systems encountered in particle mechanics (or in statistical mechanics) the integral invariant is usually called the "phase-volume" of the system.

One can consider the integral invariant $V[G]$ as an invariant measure of the set G. Equation (2) shows that this measure is nonnegative and σ-additive and Equation (3) shows that the measure is invariant with respect to motions of the system.

Very often it is convenient to investigate dynamical systems by means of spaces of functions defined on the manifold $\mathfrak{M}$. Let us consider as an example the space $L_2(\mathfrak{M})$ of functions $\varphi(x)$ which are square integrable with respect to the measure μ defined by the integral invariant of the system. The motions Q_t of the system induce a representation by means of transformations Q^t of the functions belonging to $L_2(\mathfrak{M})$ according to the formula

$$Q^t \varphi(x) = \varphi(Q_t x).$$

This defines a one-parameter group of unitary transformations of the space $L_2(\mathfrak{M})$. The group property follows from the definition of Q^t and the unitarity is a consequence of the preservation of the $L_2(\mathfrak{M})$-norm, due to the invariance of the measure:

$$\int \varphi^2(Q_t y)F(y)\,dy = \int \varphi^2(y)F(y)\,dy,$$

By Stone's theorem[21] any such one parameter group of unitary operators is generated by a fixed self-adjoint operator A:

$$Q^t = e^{itA},$$

or, in terms of the spectral resolution E_λ of A

$$Q^t\varphi = \int e^{it\lambda}\,dE_\lambda\,\varphi.$$

The operator A is called the generator (or infinitesimal operator) of the motion and is defined by

$$iA\varphi = \lim_{t\to 0} \frac{Q^t\varphi - \varphi}{t}.$$

In the case under consideration A will obviously be a differential operator of first order:

$$iA\varphi = \lim_{t\to 0} \frac{Q^t\varphi(y) - \varphi(y)}{t} = \frac{d}{dt}\varphi(Q_t y)\Big|_{t=0}$$
$$= \sum \frac{\partial\varphi}{\partial y_{t,i}} \frac{dy_{t,i}}{dt}\Big|_{t=0} = \sum Y_i \frac{\partial\varphi}{\partial y_i},$$

i.e.,

$$iA = \sum_{i=1}^{n} Y_i(y_1, \ldots, y_n) \frac{\partial}{\partial y_i}.$$

We further assume that there exists a (perfect) nuclear space Φ which is a dense subspace of the space $L_2(\mathfrak{M})$. We can then use the fundamental theorem of Section 4, which will associate to the spectral family of the operator A a complete system of derivatives

$$\chi_\lambda = \frac{dE_\lambda e}{d\sigma_\lambda} \qquad \left(\sigma_\lambda(P) = \int_P d(E_\lambda e, e)\right),$$

which are defined as linear functionals over the space Φ.

[21] See F. Riesz and B. Szökefalvy-Nagy, p. 137. (Ref. 3, this book, p. 169.)

The generalized functions f on the space Φ are in a natural way subjected to "motions" Q^t (by transposing the operator Q_t to the test function[22])

$$(Q^t f, \varphi) = (f(Q_t x), \varphi(x)) = (f(x), \varphi(Q_t^{-1} x)).$$

Applying the operator Q^t to the generalized function $\chi_\lambda(x)$, this function will be multiplied by the numerical factor $e^{it\lambda}$. Indeed:

$$\begin{aligned}(Q^t \chi_\lambda(x), \varphi) &= (\chi_\lambda(x), \varphi(Q_t^{-1} x)) \\ &= \lim_{\sigma(\Delta)\to 0} \left(\frac{E(\Delta)e}{\sigma(\Delta)}, \varphi(Q_t^{-1} x)\right) \\ &= \lim \left(\frac{Q_t E(\Delta)e}{\sigma(\Delta)}, \varphi(x)\right) \\ &= \lim \left(\frac{1}{\sigma(\Delta)} \int_\lambda^{\lambda+\Delta\lambda} e^{it\xi}\, dE_\xi\, e, \varphi(x)\right) \\ &= (e^{it\lambda}\chi_\lambda(x), \varphi(x)).\end{aligned}$$

Consequently *the generalized functions* $\chi_\lambda(x)$ *are generalized eigenfunctions of the operators of "motion"* Q^t *belonging to the eigenvalue* $e^{it\lambda}$. For almost all points λ of the spectrum there exist functionals (χ_λ, φ) defined on $D_{(1)}$ (the space of functions admitting the derivative $\partial^n/\partial x_1 \cdots \partial x_n$ such that

$$(\chi_\lambda, Q^t\varphi) = e^{it\lambda}(\chi_\lambda, \varphi).$$

For $\lambda = 0$ this would yield an invariant functional analogous to the invariant measure: an "invariant distribution." Examples of dynamical systems with "countably-multiple Lebesgue spectrum"[23] are known. For such systems one can prove that for almost every λ there exist countably many invariant functionals $(\chi_\lambda^{(n)}, \varphi)$. The problem of existence of the derivative $dE_\lambda e/d\sigma_\lambda$ at $\lambda = 0$ is not yet solved. If it were solved in the affirmative, it would follow, for example, that in addition to the

[22] *Translator's note:* This is an example of a "dual-system representation" of a group. Cf. also the discussion of such non-unitary representations on pairs of vector spaces in duality in Volume 5, and in G. W. Mackey, *Bull. Amer. Math. Soc.* **69**, 628–686 (1963), especially Section 8.

[23] A spectrum is said to be a *Lebesgue spectrum* if each spectral measure $\sigma_\alpha(\lambda)$ is absolutely continuous and all these measures are equivalent to the ordinary Lebesgue measure.

invariant measure, a dynamical system with countably multiple Lebesgue spectrum admits countably many invariant functionals defined over twice-differentiable functions.

I.M.Gel'fand and S.V. Fomin have been able to prove that dynamical systems defined in the usual manner on Riemannian manifolds of constant negative curvature belong to this class.

NOTES AND REFERENCES

Chapter I

Spaces of type W have been introduced and analyzed by B. L. Gurevich [25]. These spaces not only represent generalizations of the spaces K_p, Z^p and $Z_p{}^p$ which had been introduced previously by I. M. Gel'fand and G. E. Shilov [22], but also lead to improvements in the definitions and proofs.

A slightly more general class of spaces, together with their duality theory is due to L. Hörmander [30] (see also Additional References [1]).

B. Ya. Levin's theorem on the existence of an entire function with given generalized growth indicatrix can be found in Ref. [39], Chapter 2.

Chapter II

Sections 1–5. Holmgren's method is treated in [28]. The first general theorems on uniqueness classes for Cauchy problems for equations of the evolution type,[1] with constant or t-dependent coefficients are due to I. G. Petrovskiĭ [45]. He has shown that the class of all functions which are continuous and bounded for $-\infty < x_j < \infty$ is a uniqueness class for each of these systems. This theorem has been extended by V. E. Lyantse [41] and L. Schwartz [50] to the class of all functions of power-law growth, Schwartz making use of the theory of distributions (generalized functions over S). The construction of the uniqueness classes consisting of exponentially increasing functions for arbitrary Petrovskiĭ-correct systems is due to the authors: in Ref. [22] it was derived by making use of the Fourier transforms of exponentially increasing functions and in Ref. [23] by means of the operator method, considering operators $\exp(tp(d/dx))$ in the spaces $S_\alpha{}^\beta$. The concept of reduced order of a system was first introduced in [22], in terms of the order of the resolvent matrix $e^{tP(s)}$.

[1] That is, systems of equations of the form

$$\frac{\partial u(x, t)}{\partial t} = \sum_{|q| \leqslant m} a_q(x, t)\, D_x{}^q u(x, t) + b(x, t).$$

The characterization of uniqueness classes given in this book is more complete than in Ref. [22] and [23] (the exponent $p_0' - \epsilon$ is replaced by p_0'). This improvement is due independently to K. I. Babenko [2], B. L. Gurevich [25] and S. D. Éĭdelman [14]. G. N. Zolotarev [56] is the author of the theorem that the functions of exponential order $\leqslant p_0' + \epsilon$, $\epsilon > 0$ arbitrary, do not form a uniqueness class for any system of the evolution type (for $p_0' > 1$).

The following authors have contributed to the problem of determining the uniqueness classes for the Cauchy problem: E. Holmgren [29] and A. N. Tikhonov (Tychonoff) [52] for the heat equation and related equations, O. A. Ladyzhenskaya [38] for general parabolic equations and S. D. Éĭdelman [12] for general parabolic systems. Necessary and sufficient conditions for functions satisfying an inequality of the type $|f(x)| \leqslant Ce^{\Phi(x)}$ to form a uniqueness class for Cauchy problems have been established by S. Täcklind [51] for the heat equation and by G. N. Zolotarev [55] for Petrovskiĭ-parabolic systems.

The original definition of a hyperbolic system and the existence of solutions for sufficiently smooth initial functions is due to I. G. Petrovskiĭ [45]. A slightly more general definition, permitting to formulate the converse theorem (i.e., that any system with solutions for sufficiently smooth initial data is hyperbolic) has been given by L. Gårding [16] (cf. in this connection, Section 3, Chapter III). Similar theorems have been subsequently proved by L. Schwartz [50] within the framework of distribution theory. Another approach to uniqueness theorems, on the basis of Laplace transforms in t (which seems to be applicable only to the case of one space variable) is due to E. Hille (cf. Ref. [27] and further references given there).

Section 6. The reduction of the problem of determining the growth of the function $e^{tP(s)}$ in the complex *s-plane* to the investigation of the growth of the real parts of the characteristic roots of the matrix $P(s)$, by making use of the inequality (6) is due to G. E. Shilov [48]. The estimation of the coefficients of the Newton interpolation formula by means of complex integration has been taken from the book [24] of A. O. Gel'fond. The formula for the computation of the reduced order is due to V. M. Borok [6]. She has shown in the same paper that every system with integral reduced order p_0 can be reduced to a system of the form $\partial u/\partial t = P_0(i\,\partial/\partial x)u$, where the differential operator P_0 is of order not higher than p_0 (for hyperbolic systems: 1). If the reduced order is rational $p_0 = p/q$, the system can be reduced to the form $\partial^n u/\partial t^n = P_0(i\,\partial/\partial x)u$ with $n \leqslant q$ and the order of P_0 not larger than p.

Section 7. The main result of this section is due to G. E. Shilov [49].

It is given here in an improved version ($h < 1$ is replaced by $h < (2p_0)'$), which was obtained by I. I. Shulishova in her Diploma Thesis (Master's Thesis).

The note at the end originated with Yu. A. Dement'ev.

The results of Appendix 1 are due to B. L. Gurevich.

The results contained in Appendix 2 have not been published so far and are due to A. G. Kostyuchenko and G. E. Shilov.

A. G. Kostyuchenko [34] has also obtained the results presented in Appendix 3. These results are given here in a more complete form. The theorem of Éĭdelman on the solution of the Cauchy problem for systems with elliptic operators is contained in Ref. [13].

Chapter III

The first general theorems on uniqueness classes for the Cauchy problem for systems of evolution type with constant or t-dependent coefficients have been found by I. G. Petrovskiĭ [45]. He has shown that "condition A" (cf. p. 107; this condition is our condition of Petrovskiĭ-correctness) is necessary and sufficient in order that the class of functions which are bounded together with their derivatives up to a certain order for $-\infty < x_j < \infty$, form a uniqueness class for the Cauchy problem for systems of the form $\partial u/\partial t = P(i\,\partial/\partial x)u$. The following authors have indicated correctness classes consisting of functions of exponential type: Holmgren [29] and Tikhonov [52] for the heat equation, Täcklind [51] for the equation $\partial u/\partial t = \partial^{2p}u/\partial x^{2p}$, Ladyzhenskaya [38] for general parabolic equations, and Éĭdelman [12] for Petrovskiĭ-parabolic systems. The general construction of correctness classes for arbitrary systems of evolution type have been carried out by Shilov [48]. The presentation in this volume is the first systematic exposition of the subject.

Section 2. Petrovskiĭ-parabolic systems were first introduced in Ref. [45]. For the general definition cf. the paper by Shilov [48].

Characteristics for systems with one space variable have been computed by V. M. Borok [7]. Petrovskiĭ-parabolic systems with coefficients depending on the space variables have been investigated by Éĭdelman [13].

Since the fundamental solution of a parabolic system is an infinitely differentiable function of x, each solution belonging to a uniqueness class has the same property, although the initial function need only be locally integrable. V. M. Borok [5] has shown that only parabolic systems have this property.

Section 3. Systems which are Petrovskiĭ-hyperbolic were investigated in Ref. [45] and by Gårding [16].

Section 4. What we call Petrovskiĭ-correct systems, Petrovskiĭ himself called systems with "condition A." S. A. Galpern [15] has introduced a class of systems which occupy within the class of Petrovskiĭ-correct system the same place, as the Petrovskiĭ-hyperbolic equations occupy among the general hyperbolic systems. Systems of positive genus p_0 have been introduced by the authors [22] under the name "regular systems." Subsequently Kostyuchenko and Shilov [35] have proved the theorem that each such system has a solution within the class of functions of order $\exp \epsilon|x|^{p_0'}$ ($\epsilon > 0$ arbitrary). This proof is the basis of the general existence theorems established in this section. In Ref. [14] Éĭdelman has shown that certain systems in physics and mechanics belong to the class of regular systems (e.g., the equation describing sound propagation in a viscous gas, given in this section). Formulas for the computation of characteristics for systems with one space variable have been given by V. M. Borok [7]. Theorem 4 is due to Kostyuchenko and Shilov and has not previously been published.

For systems which we called "conditionally correct," V. E. Lyantse [41] has indicated a uniqueness class consisting of infinitely differentiable functions with exponential growth.

F. John [31] has arrived at the class of conditionally correct systems by different considerations. He has described those systems which admit at least one solution with a nonvanishing initial function of compact support.

The results in subsection 4.3 (correctness in the class of analytic functions) are due to Kostyuchenko [48]. This is the first detailed account of the subject. L. Ehrenpreis [11] has also investigated the solvability of systems which have entire functions as initial data.

Chapter IV

The history of the problem of eigenfunction expansions has been sketched in the preamble of this chapter. The reduction of quadratic integral forms to canonical form and the proof of the completeness of the eigenfunctions of a regular Sturm-Liouville problem (Steklov's theorem) are due to D. Hilbert [26]. The completeness of the eigenfunctions for compact operators was first proved by F. Riesz [47]. The spectral resolution of unbounded self-adjoint operators in a Hilbert space is due to J. von Neumann [43]. The problem of extension of symmetric operators, in particular, semibounded operators (theorems of von Neumann, Friedrichs, Riesz) is treated in [36] (cf. also the repeatedly quoted book by Riesz and Nagy). Eigenfunction expansions for an ordinary differential operator on a semiaxis are originally due to H. Weyl [54] (cf. also the more recent investigations of E. C. Titchmarsh [52] and B. M. Levitan [40]). For differential operators of nth order this problem has been investigated by M. G. Kreĭn [37] and K. Kodaira [32]. A. Ya. Povzner [46] has treated the problem of eigenfunction expansions for operators of the form $-\Delta u + qu$, defined in the whole space. F. I. Mautner [42] has proved an expansion theorem for general self-adjoint operators for which the resolvent is an integral operator with kernel of Carleman type. F. Browder [87] and L. Gårding [17,18] have shown that all elliptic differential operators belong to this class, consequently the expansion theorem is valid for any elliptic operator.

Section 2. Differentiation of functionals of strongly bounded variation has been considered independently first by I. M. Gel'fand [19] and then by N. Dunford [10] and B. J. Pettis [44].

Sections 3–7. The results of these sections are due to Gel'fand and Kostyuchenko [21]. F. Browder [9] has extended the fundamental theorem to the case of maximal symmetric operators (cf. in this connection the paper [33] of Kostyuchenko, where the theorems on the structure of generalized eigenfunctions have been carried over to this case). After Ref. [21] appeared, Yu. M. Berezanskiĭ [3] proposed a different way of obtaining the eigenfunction expansion in the space $L_2(R_n)$. In a communication at the 3rd Soviet-Union Mathematical Congress (June, 1956), Gårding reported that he has established a theorem on generalized eigenfunction expansions for selfadjoint differential operators in $L_2(R_n)$. Subsequently Berezanskiĭ showed that the primitive functions of the generalized eigenfunctions (in R_n) do not increase faster than $|x|^{(5/2)n+1+\epsilon}$. Kostyuchenko [33] has improved this result, replacing $|x|^{(5/2)n+1+\epsilon}$ by $x^{n/2}$.

In another paper [4], Berezanskiĭ has extended to eigenfunction expansions Bochner's theorem on the representation of functions of positive type.

The Sobolev problem has been investigated by R. A. Alexandryan [1].

The results of Section 6 are due to Kostyuchenko [33], those in Section 7 are due to Gel'fand and Kostyuchenko [21]. The theorem of Gel'fand and Fomin on dynamical systems on manifolds of constant negative curvature can be found in Ref. [20].

Translator's Note

No systematic attempt has been made to include the literature which has appeared since the original manuscript was completed (1957). A few newer books which are pertinent have been included as "Additional References." For further references the reader is directed to survey articles and the Mathematical Reviews for the past nine years.

As was already remarked, the last chapter is related to Chapter I in volume 4 (cf. the notes and references to that Chapter, especially Section 4, for additional bibliography).

BIBLIOGRAPHY

1. Aleksandryan, R. A., On the Dirichlet Problem for the Equation of a String (in Russian), *DAN SSSR* **73**, 869–872 (1950).
2. Babenko, K. I., On a New Problem of Quasi-analyticity and on the Fourier Transform of Entire Functions (in Russian), *Trudy Moskov. Mat. Obshch.* **5**, 523–542 (1956).
3. Berezanskiĭ, Yu. M., On the Eigenfunction Expansion for General Self-Adjoint Differential Operators (in Russian), *DAN SSSR* **178**, 379–382 (1956).
4. Berezanskiĭ, Yu. M., Generalization of Bochner's Theorem to Eigenfunction Expansions for Partial Differential Equations (in Russian), *DAN SSSR* **110**, 893-896 (1956).
5. Borok, V. M., On a Characteristic Property of Parabolic Systems (in Russian), *DAN SSSR* **110**, 903–905 (1956).
6. Borok, V. M., Reduction to Normal Form of a System of Linear Partial Differential Operators with Constant Coefficients (in Russian), *DAN SSSR* **115**, 13–16 (1957).
7. Borok, V. M., On systems of Linear Partial Differential Equations with Constant Coefficients (in Russian). Dissertation, Moscow, 1957.
8. Browder, F., The Eigenfunction Expansion Theorem for the General Self-Adjoint Singular Elliptic Partial Differential Operator, *Proc. Nat. Acad. Sci. U.S.* **40**, 454–467 (1954).
9. Browder, F., Eigenfunction Expansions for Formally Self-Adjoint Partial Differential Operators, *Proc. Nat. Acad. Sci. U.S.* **42**, 769–773 (1956).
10. Dunford, N., Uniformity in Linear Spaces, *Trans. Am. Math. Soc.* **44**, 305–356 (1938).
11. Ehrenpreis, L., Cauchy's Problem for Linear Partial Differential Equations with Constant Coefficients, *Proc. Nat. Acad. Sci. U.S.* **42**, 642–646 (1956).
12. Éĭdelman, S. D., Estimates for Solutions of Parabolic Systems and some Applications (in Russian), *Mat. Sbornik* **33**, 359–382 (1953).
13. Éĭdelman, S. D., On Fundamental Solutions of Parabolic Systems (in Russian), *Mat. Sbornik* **38**, 51–92 (1956).
14. Éĭdelman, S. D., On Regular and Parabolic Systems of Partial Differential Equations (in Russian), *UMN* **12**, No. 1, 254–257 (1957).
15. Gal'pern, S. A., On the Correctness of the Cauchy Problem for Systems of Linear Partial Differential Equations (in Russian), *Mat. Sbornik* **7**, 111–141 (1940).
16. Gårding, L., Linear Hyperbolic Partial Differential Equations with Constant Coefficients, *Acta Math.* **85**, 1–62 (1951).
17. Gårding, L., Eigenfunction Expansions Connected with Elliptic Differential Operators, *Proc. 2nd Scand. Math. Congr. Lund*, 1953, 44–55 (1954).
18. Gårding, L., Applications of the Theory of Direct Integrals of Hilbert Spaces to some Integral and Differential Operators. Institute for Fluid Dynamics and Applied Mathematics, Univ. of Maryland, College Park, Maryland (1954).
19. Gel'fand, I. M., Abstrakte Funktionen und Lineare Operatoren, *Mat. Sbornik* **4**, 235–286 (1938).

20. Gel'fand, I. M., and S. V. Fomin, Geodesic Flows on Manifolds of Constant Negative Curvature (in Russian), *UMN* **7**, No. 1, 118–137 (1952).
21. Gel'fand, I. M., and A. G. Kostyuchenko, On Eigenfunction Expansions for Differential and other Operators (in Russian), *DAN SSSR* **103**, 349–352 (1955).
22. Gel'fand, I. M., and G. E. Shilov, Fourier Transforms of Rapidly Increasing Functions and Problems of Uniqueness for Solutions of the Cauchy Problem (in Russian), *UMN* **8**, No. 6, 3–54 (1953).
23. Gel'fand, I. M., and G. E. Shilov, A New Method in Theorems on the Uniqueness of the Solution of the Cauchy Problem (in Russian), *DAN SSSR* **102**, 1065–1068 (1955).
24. Gel'fond, A. O., "The Calculus of Finite Differences" (in Russian) Gostekhizdat, Moscow and Leningrad, (1952).
25. Gurevich, B. L., New Types of Test Function Spaces and Spaces of Generalized Functions and the Cauchy Problem for Operator Equations (in Russian). Dissertation, Kharkov, 1956.
26. Hilbert, D., Grundzüge einer allgemeinen Theorie der linearen Integralgleichungen. Teubner, Leipzig, 1912.
27. Hille, E., Some Aspects of Cauchy's Problem, *Proc. Intern. Congr. Math. 1954, Amsterdam*, **3**, 109–116 (1956).
28. Holmgren, E., Über Systeme von linearen partiellen Differentialgleichungen, *Öfversigt Kongl. Vetensk.-Akad. Förh.* **58**, 91–103 (1901).
29. Holmgren, E., Sur les solutions quasianalytiques de l'équation de la chaleur, *Arkiv. Mat. Astron. och Fysik* **18**, No. 9 (1924).
30. Hörmander, L., La transformation de Legendre et le théorème de Paley-Wiener, *Compt. Rend.* **240**, 392–395 (1955).
31. John, F., Non-Admissible Data for Differential Equations with Constant Coefficients, *Commun. Pure Appl. Math.* **10**, 391–398 (1957).
32. Kodaira, K., On Ordinary Differential Equations of any Even Order and the Corresponding Eigenfunction Expansions, *Am. J. Math.* **74**, 502–544 (1950).
33. Kostyuchenko, A. G., On the Eigenfunctions of Self-Adjoint Operators (in Russian). Dissertation, Moscow Univ., 1957.
34. Kostyuchenko, A. G., On the Uniqueness Problem for the Solutions of the Cauchy Problem and of the Mixed Problem for Systems of Partial Differential Equations (in Russian), *DAN SSSR* **103**, 13–16 (1955).
35. Kostyuchenko, A. G., and G. E. Shilov, On the Solution of the Cauchy Problem for some Types of Systems of Linear Partial Differential Equations (in Russian), *UMN* **9**, No. 3, 141–148 (1954).
36. Kasnosel'skiĭ, M. A., and M. G. Kreĭn, Fundamental Theorems on the Extensions of Hermitian Operators and some Applications to the Theory of Orthogonal Polynomials and the Problem of Moments (in Russian), *UMN* **2**, No. 3, 60–106 (1947).
37. Kreĭn, M. G., On a General Method of Expanding Positive Definite Kernels into Elementary Products (in Russian), *DAN SSSR* **53**, 3–6 (1946).
38. Ladyzhenskaya, O. A., On the Uniqueness of the Solution of the Cauchy Problem for a Linear Parabolic Equation, *Mat. Sbornik* **27**, 175–184 (1950).
39. Levin, B. Ya., Distribution of Zeros of Entire Functions (in Russian). Gostekhizdat, Moscow, 1956.
40. Levitan, B. M., Eigenfunction Expansions for Differential Operators of Second Order (in Russian). Gostekhizdat, Moscow and Leningrad, 1950.
41. Lyantse, V. E., On the Cauchy Problem in the Domain of Functions of a Real Variable (in Russian), *Ukr. Mat. Zh.* **1**, No. 4, 42–63 (1949).

42. Mautner, F. I., On Eigenfunction Expansions, *Proc. Nat. Acad. Sci. U.S.* **39**, 49–53 (1953).
43. von Neumann, J., Allgemeine Eigenwerttheorie Hermitescher Funktionaloperatoren, *Math. Annalen* **102**, 49–131 (1929).
44. Pettis, B. J., A Note on Regular Banach Spaces, *Bull. Am. Math. Soc.* **44**, 420–428 (1938).
45. Petrovskiĭ, I. G., On the Cauchy Problem for a System of Linear Partial Differential Equations in the Realm of Nonanalytic Functions (in Russian), *Bull. Moscow State Univ. Sec. A* **1**, No. 7, 1–72 (1938).
46. Povzner, A. Ya., On the Expansion of Arbitrary Functions in Terms of Eigenfunctions of the Operator $-\Delta u + cu$ (in Russian), *Mat. Sbornik* **32**, 109–156 (1953).
47. Riesz, F., Über lineare Funktionalgleichungen, *Acta Math.* **41**, 71–98 (1917).
48. Shilov, G. E., On Correctness Conditions for the Cauchy Problem for Systems of Partial Differential Equations with Constant Coefficients (in Russian), *UMN* **10**, No. 4, 89–100 (1955).
49. Shilov, G. E., On a Theorem of the Phragmén-Lindelöf Type for a System of Partial Differential Equations with Constant Coefficients (in Russian), *Tr. Mosk. Mat. Obshch.* **5**, 353–366 (1956).
50. Schwartz, L., Les équations d'évolution liées au produit de composition, *Ann. Inst. Fourier* (*Grenoble*) **2**, 19–49 (1950).
51. Täcklind, S., Sur les classes quasianalytiques des solutions des équations aux dérivées partielles du type parabolique, *Nova Acta Soc. Sci. Uppsaliensis* (4) **10**, No. 3 (1936).
52. Tikhonov (Tychonoff) A. N., Uniqueness Theorems for the Heat Equation (in Russian), *Mat. Sbornik* **42**, 199–216 (1935).
53. Titchmarsh, E. C., "Eigenfunction Expansions Associated with Second-Order Differential Equations," Part I. Oxford Univ. Press, London, 1946.
54. Weyl, H., Über gewöhnliche lineare Differentialgleichungen mit singulären Stellen und ihre Eigenfunktionen, *Göttinger Nachrichten*, 37–64, 442–467 (1909); *Math. Annalen* **68**, 220–269 (1910).
55. Zolotarev, G. N., Necessary and Sufficient Conditions for the Uniqueness of Solutions of the Cauchy Problem for Parabolic Systems (in Russian), *Izv. Min. Vyssh. Obr. SSSR, Ser. Mat.* **1** (1958).
56. Zolotarev, G. N., On Precise Estimates of the Uniqueness Classes for Solutions of the Cauchy Problem for Systems of Linear Partial Differential Equations (in Russian). Dissertation, Moscow Univ., 1958.

Additional References

1. Hörmander, L., "Linear Partial Differential Operators." Springer Verlag; Academic Press, New York, 1963.
2. Lions, J. "Equations Différentielles Opérationelles et Problèmes aux Limites." Springer Verlag, Berlin, 1961.
3. Trèves, J. F., "Topological Vector Spaces Distributions and Kernels." Academic Press, New York (to be published).
4. Friedmann, A., "Generalized Functions and Partial Differential Equations." Prentice Hall, New York, 1962.
5. Berezanskiĭ, Yu. M., "Eigenfunction Expansions for Selfadjoint Operators" (in Russian), Naukova Dumka, Kiev, 1965.

INDEX

ISBN: 978-1-4704-2885-3 (Set)
ISBN: 978-1-4704-2661-3 (Vol. 3)
9 781470 426613
CHEL/379.H